BEGINNERS GUIDE TO
BIOINFORMATICS FOR HIGH
THROUGHPUT SEQUENCING

BEGINNERS GUIDE TO BIOINFORMATICS FOR HIGH THROUGHPUT SEQUENCING

Eric Lee

National University of Singapore

T W Tan

National Supercomputing Centre, Singapore
A*STAR Computational Resource Centre, Singapore
National University of Singapore

World Scientific

NEW JERSEY · LONDON · SINGAPORE · BEIJING · SHANGHAI · HONG KONG · TAIPEI · CHENNAI · TOKYO

Published by

World Scientific Publishing Co. Pte. Ltd.

5 Toh Tuck Link, Singapore 596224

USA office: 27 Warren Street, Suite 401-402, Hackensack, NJ 07601

UK office: 57 Shelton Street, Covent Garden, London WC2H 9HE

Library of Congress Cataloging-in-Publication Data
Names: Lee, Eric Cheng-Yu, 1984–   author. | Tan, Tin Wee, author.
Title: Beginners guide to bioinformatics for high throughput sequencing / Eric Lee,
    National University of Singapore, T.W. Tan, National Supercomputing Centre, Singapore,
    A*STAR Computational Resource Centre, Singapore, National University of Singapore.
Description: New Jersey : World Scientific, [2018]
Identifiers: LCCN 2017054502| ISBN 9789813230514 (hardcover : alk. paper) |
    ISBN 9789813231665 (paperback : alk. paper)
Subjects: LCSH: Bioinformatics. | Biology--Data processing. | Genomes--Analysis.
Classification: LCC QH324.2 .L 2018 | DDC 570.285--dc23
LC record available at https://lccn.loc.gov/2017054502

British Library Cataloguing-in-Publication Data
A catalogue record for this book is available from the British Library.

For any available supplementary material, please visit
https://www.worldscientific.com/worldscibooks/10.1142/10720#t=suppl

Preface

Who is This Book For

If you are a beginner to the field of bioinformatics, or a biologist thinking of entering into the field of genomics and genome sequencing, and realize immediately that your training does not equip you to manage the computational aspects of next generation high throughput sequencing, then this book is ideal for you as a gentle introduction.

Those who are in the field of genome sequencing and already familiar with the procedures of analysis, may also find this book useful in closing some knowledge gaps here and there. You will find this book a breeze to read, and perhaps some suggestions in this book new to you, may be something you want to try out.

Questions

We truly value your comments and feedback where we have not explained something clearly. Please do send your queries to book@apbionet.org.

Eric Lee
Tan Tin Wee*
May 2018

*In many Eastern cultures, especially Chinese, the convention is for family name to be written first, as in Tan being the family or surname and Tin Wee as the given name (often joined in mainland China or hyphenated in Taiwan, or written as two separate words in Singapore, where the second author hails from), unless written in a westernized way, like Lee as in Eric Lee.

Translated Preface of the Chinese Edition "Getting Started with Next Generation Sequencing"

by C-Y Lee (Eric Lee)

Ever since the digitalization of biological data, information technology and information science have been used to solve biological problems. This approach, now known collectively as Bioinformatics, is characterized today by harnessing the power of computers, including and especially the personal computer, to process data that has been previously manually handled, in a faster and automated manner, to generate results that scientific researchers can use to figure out the biological meaning of living things.

This new and emerging discipline, combines and integrates applied mathematics, computer science and statistics in studying biology. As a research field, its scope is quite extensive and rather broad, ranging from areas such as protein structure prediction, sequence alignment, sequence assembly, gene expression and genetic evolution and so on.

Researchers who enter the field of bioinformatics logically follow their specialty and interest, based on whether their background and original expertise is either in biology or computer science. For instance, if one is a computer scientist, perhaps one may be interested in applying one's understanding of algorithms, and deploying machine learning in synergy with statistical theory, to generate from the original raw data some meaningful computed data output; or if one's foundational capability in programming is adequately strong, with algorithms and data structure theory, one could design new bioinformatics tools to solve specific problems and to achieve faster and more computationally efficient procedures that conserve system resources.

If, however, you were previously a biology student, what you will naturally do is to analyze the results of a computational analysis from a biological perspective to infer its significance. And then, through the use of biological experimental methods, you would verify if the predictions produced by the computational results make sense. A computational calculation, without validation by biological experimentation, at best, can only be regarded as prediction. Frankly, to be realistic, without experimental verification of the computed results, there is no way to publish them in journal papers. All in all, biology and computing in the context of their usage and application in bioinformatics are mutually reinforcing and need to go hand in hand.

As long as you can generate results through computing, Bioinformatics, as an experimental discipline, does not need to be kept locked up in the laboratory. Sadly, there are too many non-IT savvy professional researchers, who surrender at first encounter with a piece of research that needs computing and computational analysis.

In fact, as long as the operational capabilities of the software fulfill the basic requirements of the research study, especially under the Linux text-based command line interface, you can thereafter follow the same methodology, using the operation of different software tools, to complete your computational experiments. There is therefore really no need for us to be overly anxious about this process. We should merely treat our own operational skills in computing with confidence.

From the first two chapters, I will be familiarizing you with the computer, starting from a discussion of its very basic concepts and its underlying operations. In particular, I will describe the usage of the Linux operating system and explain its basic commands. Thereafter, using some of these commands, we will go through a number of bioinformatics tools. My basic principle is to teach you the minimum you need to know, and even then, only what you will frequently use. Difficult skills or advanced operations, which biological researchers and Linux beginners do not need, will not be covered. In fact, simply mastering these basic skills and concepts will already propel you miles ahead.

Chapters 3 and 4 of this book cover the basic operation of bioinformatics tools, narrowing it down to only "high throughout sequencing." Sequence quality inspection, alignment and comparison will definitely be encountered in any research workflow in this area. You need to understand your raw data, and then, from sequence alignment launch into other analytical tools. First, I will gently explain in simple terms what high throughput sequencing is all about. This, to a bioinformatics newcomer with a computer background , is most important. After that, regardless of whether you are a researcher from a biology or computing background, you really do need to master Linux commands, from introducing you to the simplest, uncomplicated parameters for operating a command, and gradually guiding you to attempt more variations. You will learn currently the state-of-the-art tools, which most bioinformatics researchers are using, including FastQC, FastX-Toolkit, Bowtie, SAMtools, etc. From the practical usage of these sets of tools, you will be totally familiar with the philosophy of their form and usage.

Although Chapter 5 brings you into graphics processing unit (GPU) card technology, my intention is to introduce you to the latest in computing technology. Through a detailed description of the underlying operation of a GPU processor, I hope the beginner will encounter the problem of the operating system level settings and learn how to deal with it. The GPU card is operated by CUDA, which in the past couple of years have become extremely popular and adopted as the direction for high performance computing. In the realm of bioinformatics, there are also a number of practical GPU applications. However hot this area may be, it does not mean that the technology is mature. But being immature does not mean it has less potential to become a dominant player. That is why I am duty-bound to inform the student of bioinformatics its current problems, and the industry's active progress and actual deliverables in this field. By paying close attention to the rapid development of the graphics coprocessor, bioinformatics researchers will be expected to attain greater acceleration in the speed of computation of their datasets.

Until Chapter 6, my book would have given you sufficient technical ability. However, for good results, the most substantive contribution would be to plan for the publication of your work, that is, to take your research workflow to the next level, and subject your work to the scrutiny of external peer review. In this chapter, I will explain how you can take individual commands and assemble them into a series of shell scripts, which can be processed as a complete single program. Given that many researchers with a biological background do not have the experience of writing a series of commands into a single process, I will share some simple problem solving tips, including how to write programs with minimum standards of maintainability and readability. We will also analyze two practical applications, with reference to commonly occurring pitfalls, which newbies are likely to commit. In this way, I hope that you can avoid setbacks in your research process and assemble an effective analytical workflow.

We hear a lot about cloud computing these days, and needless for me to say, it is yet another important technology we should consider for the future development of bioinformatics. In Chapter 7, besides the previously discussed two "industry-standard" practical applications, and a "commercial cloud exemplar" use case, the most important aspect we will consider is the Galaxy cloud computing platform. From the basic operation to deeper know-how of the comprehensive Galaxy bio-computational experimental pipelines and workflow, to the installation of a Galaxy server, I have deliberately chosen to discuss this last because the Galaxy bioinformatics platform for beginners is a double-edged sword. Once you have tasted the effectiveness of Galaxy, in order to avoid the frequent errors which you get from trying to execute Linux commands from memory, you will increasingly develop a heavy and perhaps unhealthy reliance on Galaxy, for many bioinformatics tools still continue to rely on the traditional command line interface. But mastering the Galaxy interface is definitely not a bad thing, because through a form-filling interface for command input, the chance of errors can be dramatically reduced. Now that tablet computing is so popular and pervasive these days, the use of the built-in browser operations forms a very practical approach towards the deployment of bioinformatics tools for research.

Finally, I wish to remind everyone to take a serious look at the Appendix. It is very important because some of what I consider to be very important content, which could not make a full self-contained chapter, is assembled here. This is especially useful for graduate students, particularly the kind of information no one will voluntarily share with you.

I hope that through this effort to share some of my experiences, the journey undertaken by you, as a bioinformatics novice, will proceed smoothly. The pressure you constantly face, in having to prepare reports periodically and to decide on what can be suitably reported in your research publication, is definitely familiar to me. Read the Appendix to find some helpful documentation on the pipeline approach towards solving these issues.

Bioinformatics data processing can be filled with techniques which are very complex. But you can also find simple ways to achieve the same goals. In this book, I chose to introduce bioinformaics in next generation sequencing using a simple approach. When all is said and done, my aim is to take care of the majority of our dear readers of the book who do not have much experience in this new field. After all, for those who just wish to become master users of regular expressions and so on, ordinary users of office productivity tools could just as easily use Excel!

Eric CY Lee
2012

Contents

CHAPTER 1

Preparing Your Computing Environment

1.1 Buying Your Own Computer

As a bioinformatics researcher, it is perfectly reasonable to wish for an ideal computer as a necessary computing tool for your work. In reality, the main purpose of your computer may merely be to serve as a terminal, because for bioinformatics these days, due to rapid progress in information technology and the development of biotechnology, the volume of data is so huge that most computational work can no longer be done on a PC — one has to rely on computing on supercomputers. In actual routine situations, if sequence assembly has to be carried out on large servers running jobs continuously for more than several hours, even extending to days and weeks of computing time, how can we expect a personal computer to be able to do the job?

Perhaps for these reasons, in selecting a functional computing platform with not too exceptional performance, the most important consideration is that of stability of the terminal interface during run-time. Suppose you were picking and purchasing a personal computer for practical usage (in a standalone system), we

1

would recommend buying one with a larger disk storage and more memory. A larger hard disk storage is needed because you may need to temporarily use your own computer to store raw data and analyse the results, and you need more memory on the basis that you need speedy analysis. What we think would be useful is that you should be able to use Excel to open the results file and carry out statistical analyses as the next step, or export it to a free open source statistical platform such as R.

A further enhanced expectation would be to opt for a 64-bit multicore computer, because there are some bioinformatics applications that only work on a 64-bit environment. Such a requirement for a personal computer is very much attainable, but to achieve an environment supporting a 64-bit architecture without installing a 64-bit operating system is also pretty futile.

In practical terms, it has been mentioned earlier that most calculations are completed on the server, which is usually also the experimental "infrastructure", that establishes an internal open shared computing environment. If you are not a computer hardware *afficionado*, then forget about this, as maybe asking you to immediately install Linux is impossible. Given the focus of this section, I would like to leave it to the chapters and sections detailing servers.

However, if you still want to have a computer supporting a 64-bit operating system, you are recommended to buy an Apple Mac series computer directly if you do not already have a mental block against non-Windows operating systems. The Apple Company is now producing machine types that are 64-bit multi-core hardware architecture, with a preloaded UNIX architecture, FreeBSD, as the basic foundation of its macOS (formerly named OS X) operating system.

After booting, you can directly have a complete environment to learn Unix shell commands to create a remote connection with the server directly in this test; there are no problems. Therefore, many bioinformatics software are native Mac applications. Its mode of operation, including a graphical interface, in terms of performance is better than the Windows version; the command operations are more humane than Linux versions. This is a personal practical experience to share with you. We really believe that spending too much time on the system used in order to establish a working environment is very wasteful and meaningless.

Today's generation is a cloud computing generation. Besides purchasing a bare metal machine as we described above, "Infrastructure as a Service" (IaaS) is your new choice. Just pay the usage fees for your actual usage time, and you can have a private computing resource directly accessible from any computer anywhere, or even from a mobile device. We did a quick survey for general IaaS providers when we updated the book, and found that the fees start from USD$5 per month. If you

Figure 1.1   *In today's retail shops, whether a desktop PC or a notebook computer, there will be 64-bit systems (for example, Intel Core i5 4-processor notebook computer). There is also the option of getting a GPU graphics processor capability in your PC (notwithstanding the crucial issue of whether one can use them properly). Running basic parallel computing on such a platform is not a problem, and using it as a dumb terminal would be more than sufficient*

just want to practice the Unix shell command, the low-end level service is enough. If you are at an advanced stage, you can pay a higher fee for a more powerful dedicated server with optimized memory, solid state drive and GPU. The flexibility is tremendous, compared to buying your own bare metal machine for your home or office.

This kind of cloud service usually provides a control panel via a typical Web browser interface such that you can "create" a machine with the desired installed operating system in seconds. With the snapshot feature, rolling-back to the initial state is simple. So not to worry about any minor mistakes, you can easily undo any damage you accidentally create.

Otherwise, you may apply for publicly available computing resources from your government or research institution. Sometimes it's free.

Note that whether you opt for cloud services or institutional resources, you will still need your own computer or device to access them. In fact, many of us are increasingly using our mobile devices, whether it is a smart phone or a tablet computer, to access these powerful and flexible remote computers and reducing the need for a desktop.

1.2   Setting up a Computing Server

After being introduced to how to choose and purchase a standalone PC or to use remote computing resources, perhaps some of you may have already acquired a certain standard of conceptual understanding in these matters. Some may already be familiar with the Linux operating system, or consider yourself a power user or even UNIX guru. It may just be that you are not too familiar with the environment of bioinformatics applications and need additional affirmation. Or perhaps you have reached a certain level of competence that you have been assigned responsibility for system administration and maintenance. We would like to share with you in this chapter some of our experience with Linux servers in the bioinformatics applications space.

In fact, various flavours of the Linux system which you will typically encounter, e.g. Linux distributions (or distros as they are fondly called) such as Ubuntu, Fedora, Red Hat, CentOS, Debian, actually originate from a common base, with most commands, functionalities and usage being the same as the Linux kernel itself as a unified specification. The only difference lies in the development kit, where developers have acquired their own taste and evolved their own design philosophies and architectural preferences. In fact, as long as system administrators and IT managers pick a familiar and comfortable flavour to use, it should suffice.

The only caveat that remains is still the end-user application requirements. In particular, the obvious limitation is that of the operating system environment. Recall that the basic entry requirements for a server is a 64-bit operating system, while the minimum requirements for memory is also very important. Prior to installing the software, one should look at the relevant specification requirements, which a subsequent chapter on software operations will explain how to assess such criteria.

Back to the limitations of the software itself, should one pay attention to whether the necessary supporting software libraries are already pre-installed? This is because most software development builds on and inherits a pre-existing software framework structure or prior completed subroutine modules. Software libraries organize the framework for development such that whenever the need arises, a call to the library can be conveniently made, thereby cutting down the time needed for the software developers to debug or troubleshoot their programs. So what types of library support are there? One popular run-time library is the Java Runtime Environment, also known as JRE, which you should install. Look also at the version number of the software, noting their backward compatibility, so you can install updates right up to the latest version.

Next, let us look at the actual software installation process, which is usually a point of confusion with newbies. Today's smart phone users are familiar with having to go to Google Play or the Apple App Store to search for and install apps from the respective app repository with the installation process made so simple and almost idiot-proof. Similarly, every Linux software installation package has instructions that are not necessarily the same. For example, Ubuntu uses `apt-get` to install from its software repository, and for Fedora, we must use `rpm` instead. For JRE, whether it is the Oracle or the Open source version, if we use the wrong version of the software package, then nothing works. This diversity of software packaging management system is a common point of great confusion among many beginners.

Next, regarding Linux system maintenance, you may wonder why Ubuntu or Debian both use "`apt-get`" for software installation, while Fedora, RHEL (Red Hat Enterprise Linux) and SUSE use "`rpm`"? The reason is because Ubuntu and Debian are of the same genre of Linuxes, and use .deb format with tools such as `apt`, `apt-get` and `dpkg`, whereas CentOS, SUSE, Fedora, and othes of the Red Hat family use .rpm format and the corresponding installation tools such as `yum` or `dnf`.

Looks like we have gotten everyone confused by now. Our opinion, which you may disagree with, and recommendation for newcomers to the open source Linux is to install the free Ubuntu flavour. It is among the more regularly updated, well-supported and most comprehensive Linux flavours, with on-time releases of major updates twice a year. In addition, it has a "Long Term Service" (LTS) version, which as a major requirement for enterprises, has regular official releases which are supported on the desktop and enterprise server versions for 3 and 5 years respectively, free of charge! After version 12, LTS will have 5 year support for both! This means that if you are a graduate student installing the latest version of Ubuntu LTS upon starting your degree, you do not really have to worry about updates and support for the next four years at least!

Table 1.1   *Linux distributions pre-installed with bioinformatics software*

Name	URL	Based system
BioSLAX	https://bioslax.apbionet.org	SLAX
DNALinux	http://www.dnalinux.com	SLAX
BioconductorBuntu	http://bioinf.nuigalway.ie/bioconductorbuntu.html	Ubuntu
Bio-Linux	http://environmentalomics.org/bio-linux/	Ubuntu

Besides, there are also many Linux environments developed especially for bio-informatics work, such as Bio-Linux. In keeping with the cloud computing times, there is even a CloudBioLinux (www.cloudbiolinux.org) released by the well-funded NERC Environmental Bioinformatics Centre, UK. These bioinformatics Linuxes are pre-installed with all major bioinformatics tools, complete with com-prehensive guides and documentation. All the usual file processing utilities, con-version software and viewing tools are conveniently included. We recommend this version of a bioinformatics platform because it is based on Ubuntu LTS which provides the necessary stability and reliability, particularly for beginners who are not going to be mucking around with updates and installations, other than the simple and easy to use apt-get for new software updates.

Finally, one important thing is that if you intend somewhere down the road to use graphics processor units (GPUs), then the choice of Ubuntu is ideal, because Ubuntu has strong support for graphics card operations, which are increasingly used to accelerate practical applications in bioinformatics and a subsequent chap-ter will elaborate on this.

1.3 Establishing a Remote Connection to a Server

In the next chapter, we will be learning about UNIX commands, and for this, we need you to be able to access a command line through a terminal window or the "console" (or shell, or command line prompt, they all mean the same thing). If you are already using a macOS, or a Linux operating system on your standalone com-puter, say, Ubuntu with a Graphics User Interface windows manager, like Gnome or KDE, then you can immediately call the terminal emulation software (typically called Console or Terminal) to give you a command line prompt in a window. If not, or if you need to connect to a remote computer server which your work or school offers, then you will need to learn how to establish a remote connection.

Before you can establish a remote connection, please contact your system administrator or lab seniors to find out how your machine connects to the outside world via the Internet. You will typically need to ask to help you create a user account on the UNIX server you are allowed to use, and set up your password access. They will tell you the Internet Protocol (IP) address of the computer host acting as a server, or the equivalent domain name of the computer server you are accessing. Typically, you will be asking for "an ssh login to a unix server", that's the jargon. In many institutions, they will require you to have another step for security, i.e. the VPN (virtual private network), so remember to ask them about this VPN access if you need access from anywhere other than from within your office or campus network.

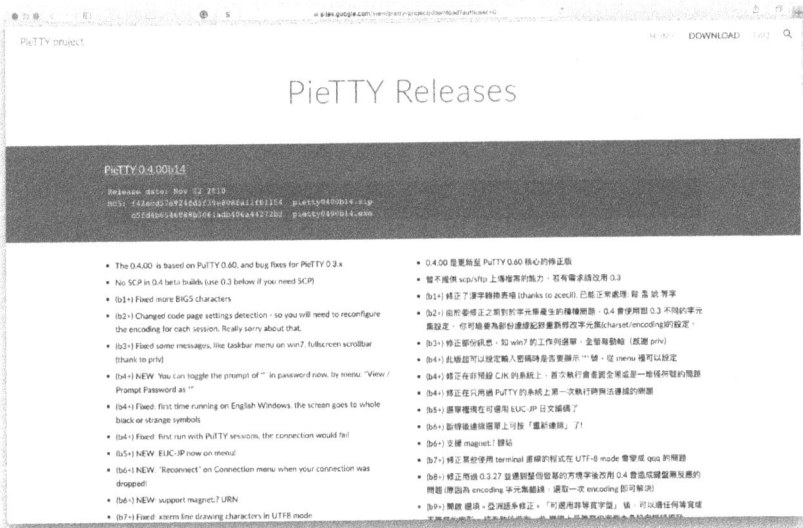

Figure 1.2   *PieTTY installation download site: http://ntu.csie.org/~piaip/pietty*

If you are using a Windows OS, please download a free software application such as PieTTY (multilingual support) or the original PuTTY from Simon Tatham. These software are free, open source terminal emulator software which creates a window for a serial console to a server, so that one can issue commands to that server for execution of any computer programs, from simple file-directory management, network file transfer, to the most sophisticated supercomputing application.

After the download, unzip and directly execute the main program. First, enter the address of the host name or IP address, and port number is by default 22. Every computer connected to the Internet, whether by WiFi or by direct cable to your local area network (LAN), will have to be configured with a valid unique Internet Protocol number, just like a telephone number on your mobile phone. This number or the IP address typically looks like 137.132.123.4, which is a series of four numbers 0 to 255, separated by dots (periods) nnn.nnn.nnn.nnn. In the near future, the new generation of IP version 6 will replace such IPv4 addresses with a much larger address space, by using IPv6 addresses which look like gibberish e.g. (2001:0db8:85a3:0000:0000:8a2e:0370:7334) and much longer at 128 bits rather than the 32 bits IPv4 addresses. Because these numbers used by computers are so long and difficult for humans to remember, there is such a thing on the Internet architecture called the domain name system which uses a memorable string of letters to represent such IP addresses, e.g. www.google.com matches 74.125.235.16 or

17 or 18, etc. So a domain name may actually connect to quite a number of different physical computers depending on how the DNS server for this domain is configured.

Even though you may know the location of a computer on the network, before you can connect to it, you need to know which port to connect to. Typically, ports are like TV channels with numbers. Each port number links to a particular computer service the server is offering to its users. So a web browser will connect by default to a web server at port 80. If you wish, a web server administrator could just as easily set up a web server at any port number but once it is not the default 80, you would have to specify the port number somewhere. Similarly, to access a server's console via a terminal emulator, one would typically connect via a specific protocol such as telnet (port 23) or the more secure SSH protocol (port 22).

Both PieTTY and PuTTY client software establish a secure connection to the host server by encrypting the transmission of data between server and client. This encrypted transmission is particularly important to prevent eavesdroppers on the Internet from intercepting your Internet communication packets and finding your passwords which can be used to break into a computer server.

What about other ports? Just like Web Port 80 or ssh at port 22, bulletin board service typically uses port 23, like the file transfer protocol (FTP) uses port 21, and simple mail transfer protocol SMTP for most of your email uses port 25. These are default port numbers assigned and set by the IANA, the Internet Assigned Numbers Authority.

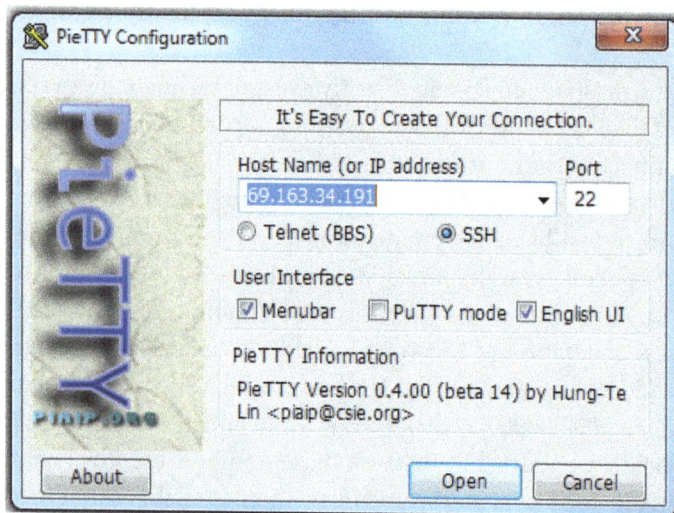

Figure 1.3 *PieTTY screen after execution of the main program*

PieTTY Security Alert

The server's host key is not cached in the registry. You have no guarantee that the server is the computer you think it is.
The server's rsa2 key fingerprint is:
ssh-rsa 1024 74:37:d7:1f:ac:64:a6:17:af:45:3d:b2:64:94:86:46
If you trust this host, hit Yes to add the key to PieTTY's cache and carry on connecting.
If you want to carry on connecting just once, without adding the key to the cache, hit No.
If you do not trust this host, hit Cancel to abandon the connection.

Yes   No   Cancel

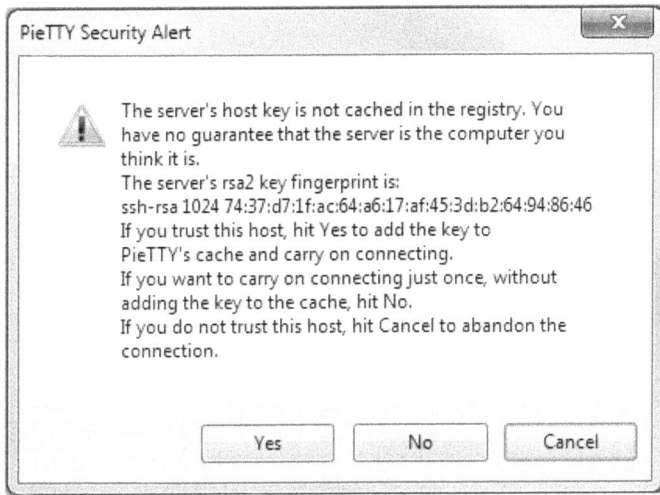

Figure 1.4 *Click "Yes" to accept the server's encryption keys. The set of keys exchanged in this manner is used for encryption and decryption purposes, to ensure a secure encrypted transmission which is less likely to be compromised*

The keys will have to be updated if there are changes to the system, e.g. if the IP address or domain name is changed, or if you reinstall the entire server. A new set of keys will have to be exchanged. Good software like PuTTY or PieTTY will give a key change warning, reminding you to be careful and to check if you are connecting to a host computer that has been hacked and compromised (if you do not know, please ask your seniors or your friendly system administrator). Assuming all goes well, and you trust the host given keys, then you will be presented with the login screen of the host. Please enter the account login user id and password in sequence and then you will see the login screen of the host with the command line prompt.

If the functionality of PieTTY is too simple and not sophisticated enough for you, we would recommend to you another terminal connection software: Xshell. This is a paid commercial software whose main powerful feature lies in the ability to page through and manage logins to multiple servers. Although it is commercial software, requiring payment, if it is used at home or at school, you can select the license type for home or school during the installation process and you can get a free licence.

Finally, we discuss the terminal login to a Linux operating system from a Mac operating system. macOS has a "Terminal", but does not use the method for Windows as described above for PieTTY or Xshell. Whereas the Windows version prompts you to enter something, in the macOS you can use Spotlight to enter

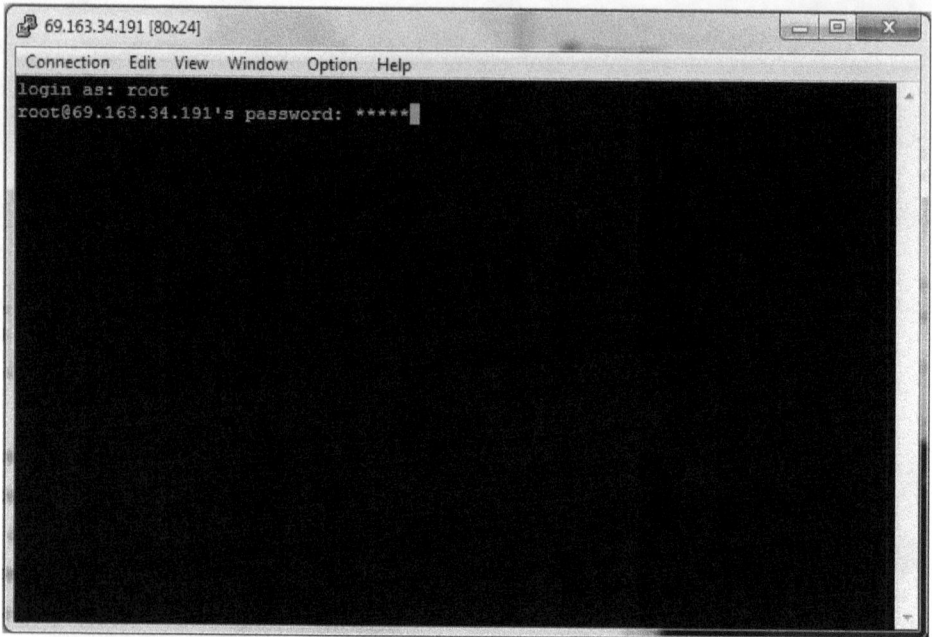

Figure 1.5 *Enter your username and password, previously obtained from your system administrator. Remember to press* ENTER *each time you finish entering the username or password*

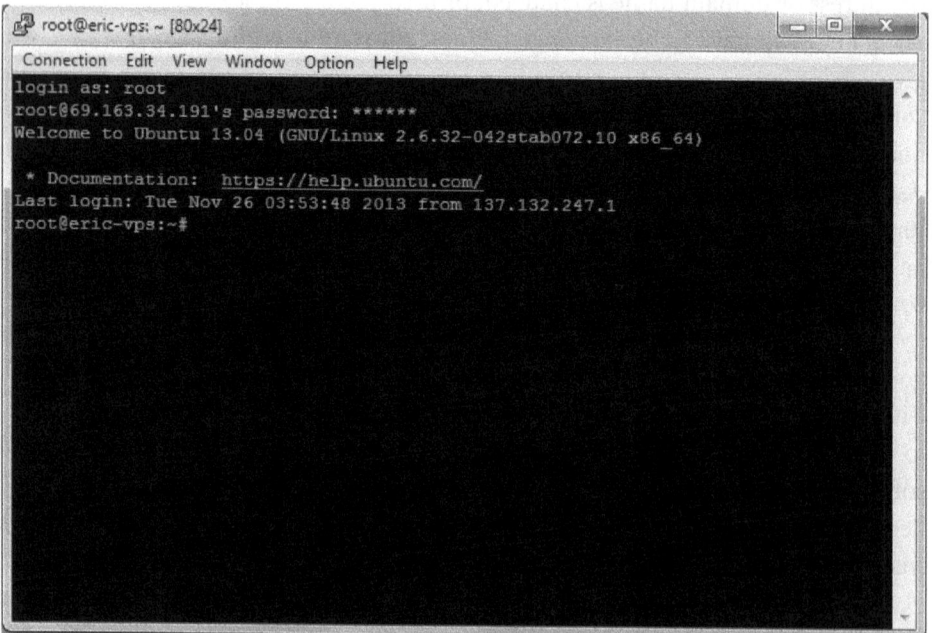

Figure 1.6 *Login procedure to the host if successfully completed, you will see the command prompt in the user input screen*

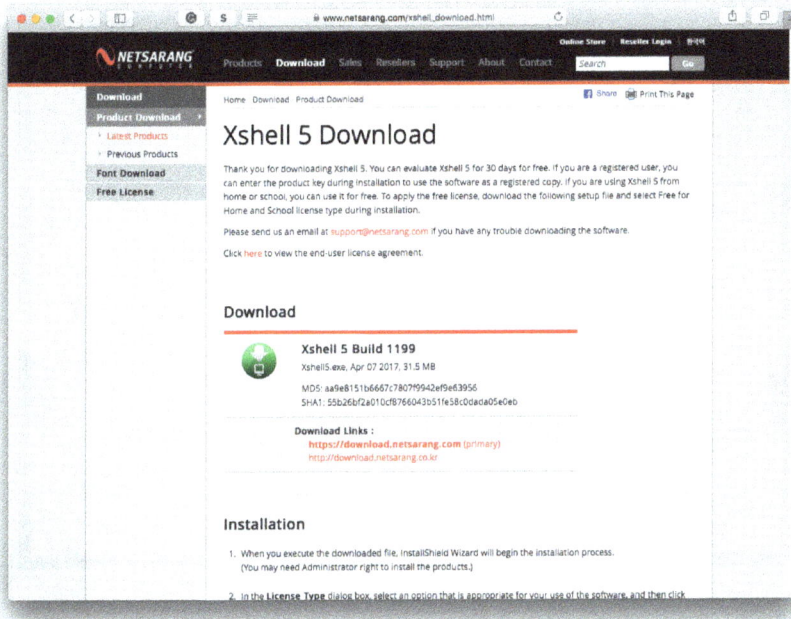

Figure 1.7 Free download of Xshell

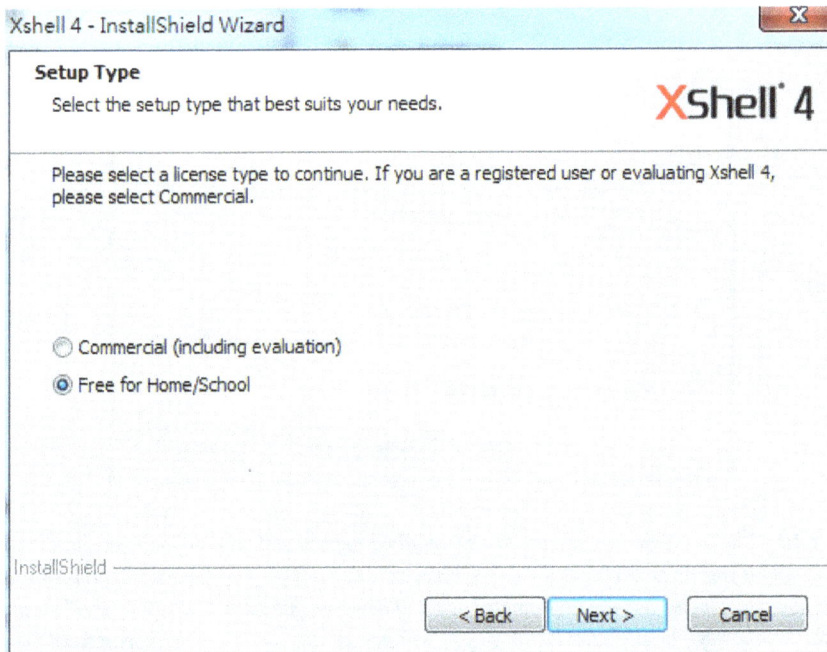

Figure 1.8 Note that at the beginning of the installation procedure, select the type of license as *Home/ School, which is free*

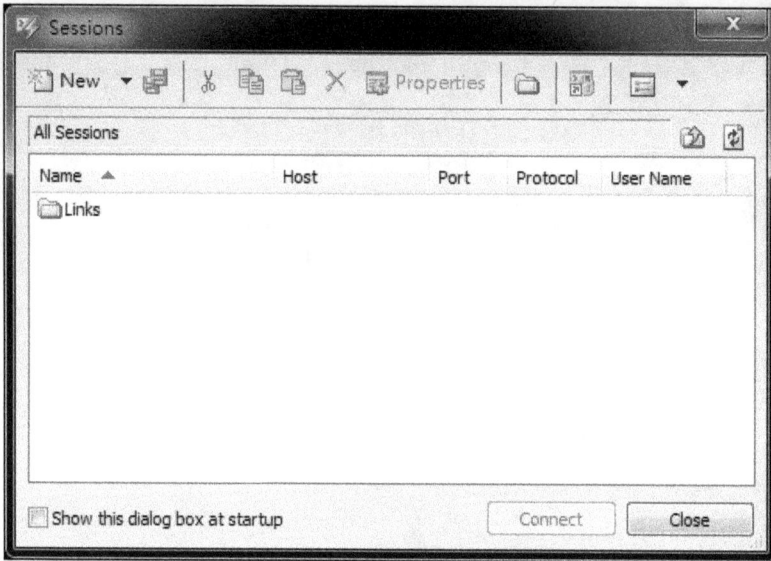

Figure 1.9 *Xshell under the Sessions tab will store and recall multiple login sessions to multiple servers, creating a very fast and convenient shortcut for connection to remote servers. Configure your sessions by keying in the IP address of your server or its domain name*

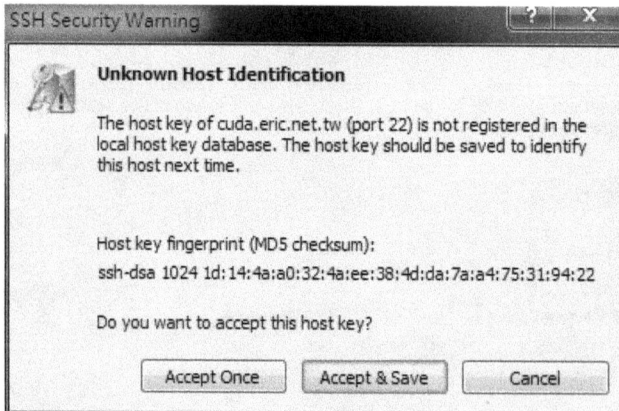

Figure 1.10 *The first time you connect to a server, the program will prompt you to accept the encryption keys, and if you select "Accept Once" it will store the key just for this session and proceed to log in. For the subsequent login, you will be prompted again. If however, you select "Accept & Save" the application will save the keys for all future logins, so you will not see this screen again, at least until the server parameters change and a fresh key needs to be exchanged*

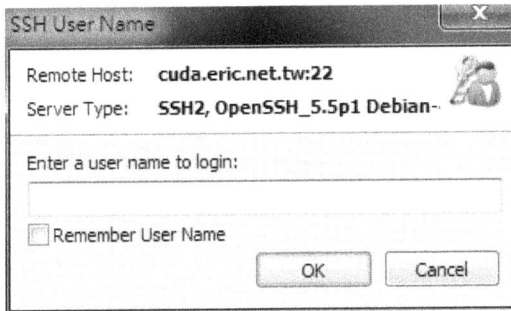

Figure 1.11 *For the first login, you need to key in your user id and click "OK"*

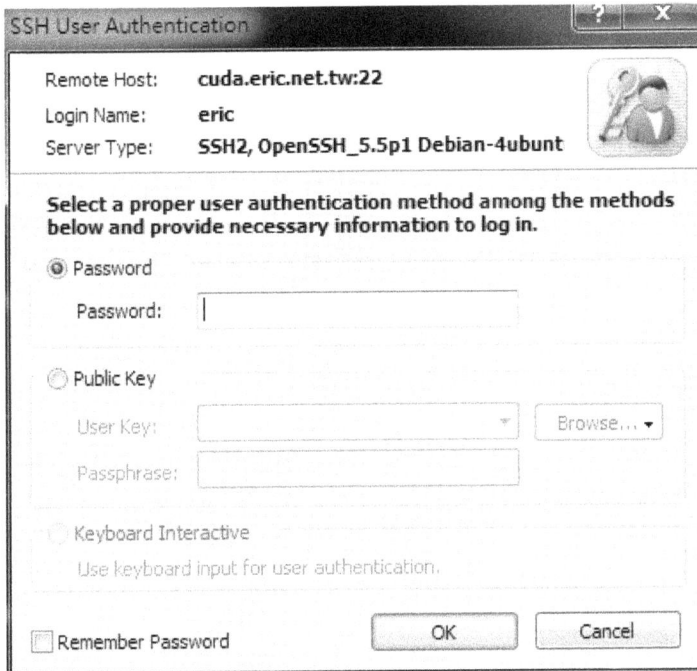

Figure 1.12 *After keying in your password, note the bottom of the window. There is a "Remember Password" check box. If you check it and click OK, all future sessions to this server will automatically log you in. We do not really recommend checking this box, because it bypasses the keying in of your password, which itself is a security risk, if someone assesses your machine without your knowledge. Moreover, by relying on the application to remember your passwords for you, you tend to forget your password once you get used to this feature*

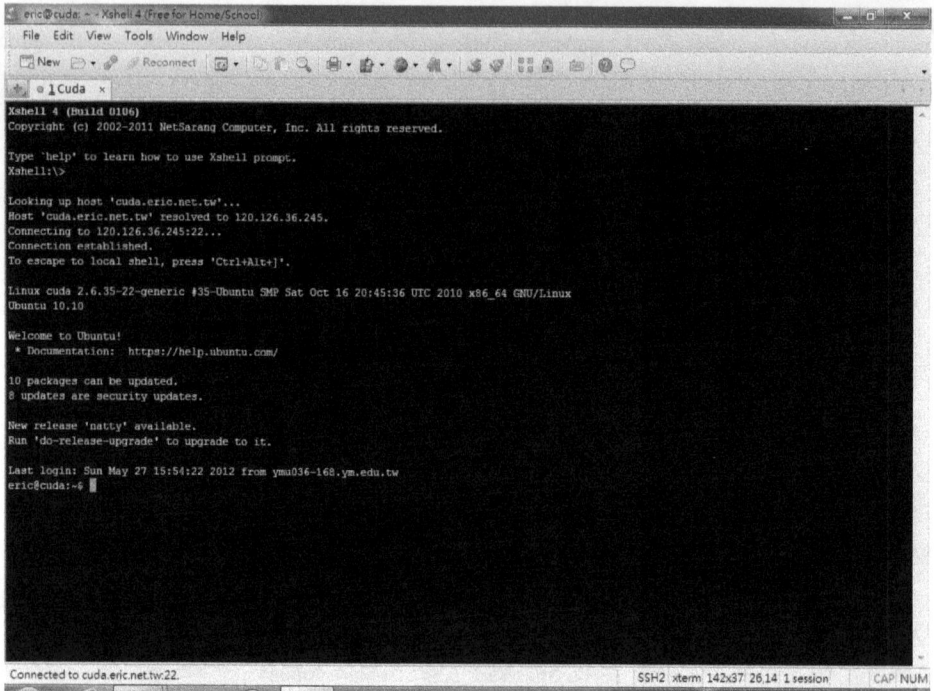

Figure 1.13 *Once logged in, both Xshell and PieTTY interfaces look very similar. (Of course we would expect it to be very similar because both are logging into the same host server!)*

Terminal, or through the Finder application (under Applications), System Tools (Utilities) to find the Terminal application. This Terminal application gives you the macOS command line prompt, which is actually a UNIX command prompt. Both Linux and macOS are derivatives of UNIX.

At the command prompt of the macOS Terminal, enter the command:

```
ssh username@server address
```

For example,

```
ssh jackiechan@zodiac.com
ssh chowyf@123.1.2.15
```

```
● ● ●              ⌂ eric — eric@cctv: ~ — ssh eric@cctv.eric.lv — 80×24
Erics-Mac-mini:~ eric$ ssh eric@cctv.eric.lv                              ▨
eric@cctv.eric.lv's password:

The programs included with the Debian GNU/Linux system are free software;
the exact distribution terms for each program are described in the
individual files in /usr/share/doc/*/copyright.

Debian GNU/Linux comes with ABSOLUTELY NO WARRANTY, to the extent
permitted by applicable law.
Last login: Sat Nov 21 09:56:04 2015 from 175.156.9.243
eric@cctv:~$ ▨
```

Figure 1.14   *macOS built-in terminal login screen*

Please note that like in all command lines, the ssh command must be separated from the argument and any parameters by blank spaces, in this case, the user name and server address. We will be discussing a lot more about Linux command line commands in the next chapter to get you used to this.

The above screenshot shows the actual login to a server from a macOS Terminal. Although the macOS terminal design is very simple, it also supports paging and terminal display styles, such as turning the Terminal window into a green on black (Homebrew) like that from the movie, Matrix.

These simple steps outlined above with almost step-by-step screenshots should be practised a few times and once you get used to it, you will be so skilled that we can move to the next chapter and show you how to enter commands and do some basic Linux command line operations.

CHAPTER 2

Learning Basic Linux Commands

2.1 No Need to be a Linux Guru to use Linux Effectively

There is already a lot of resource material and guidebooks in the bookstore or online if you wish to learn or master Linux. Personally, I would recommend a few sites, like my favourite Chinese website "Linux of Vbird"! So do scout around the net and find a blog site, a manual site, a tutorial site, or a video clip site from YouTube which reaches out to you and you immediately feel comfortable with, or buy a book like this one, which you can refer to while reaching out to a mobile device and try the Linux commands interactively. Nothing beats trying it yourself on the Linux command prompt and experimenting various options.

Few people in the world are very strong in the biological sciences, and are masters of information technology at the same time, much less be a system administrator with access to system management of a server as well. If you have the ability in both these domains, you would typically be a person with an inherently biological background, coupled with a deep interest in computers and in programming. Many bioinformatics beginners do pretty well even though they may not have touched Linux before. But for Next Generation Sequencing, which is the subject of this book, it is essential to have a good grasp of Linux. Typically for the newcomer to the interface between IT and Biology, it is not really necessary to lumber yourself with too much technical expertise, and as such, having basic understanding of Linux should be enough. If you get too embroiled in the complexities of the system, you may end up forgetting the true aim of becoming a bioinformatics expert, with strong skills in processing data emerging from next generation sequencing.

Based on the above reasons, in this chapter, we will be entirely practically oriented, so you will learn how to view files, change permissions, understand the system's configurations, deal with data compression and packing, etc., everything that you need to organize your work environment as you deal with large datasets from NGS in the form of numerous files and folders.

2.2 Folder (Directory) Operations

Folders (or Directories) exist to allow us to organize files and aggregate like files and group them into a particular location. When you login into a Linux server and arrive at a command prompt, you land into a default home folder or home directory (we will use folder and directory interchangeably). Anything within this home directory, you have full rights and permission to create new subfolders, or new files, to delete them, or to rename them. Under the Linux operating system, no other user on the server (other than the superuser or root account, owned by the system administrator) will be able to view or access your folders (and subfolders) without you giving them permission. In a subsequent section we will learn how we can fine-tune the permissions for each folder to allow groups of users or others to access your folder. Right now let us concentrate on your current folder.

```
Command: pwd
Action: find out current folder (print or path of
        working directory)
```

Before we issue any commands, let us be absolutely clear about the concept of Linux as a case sensitive operating system. Uppercase and lowercase are distinguished. So a command pwd is totally different from the command PWD. If you ignore this little detail, you will run into much trouble everywhere in the Linux environment, and end up frustrated.

Next, a space in between commands and arguments is significant, and makes a world of difference. But two or more spaces typically mean the same thing at the command prompt, or spaces at the beginning or at the end (trailing spaces) generally are harmless.

Every character means something, so if you make a mistake, and insert a superfluous character into the command line, please delete it first before you even touch the ENTER key to execute the command. Imagine, the following will actually delete all your files in your system: `rm -fr /` (don't ever try this at home), and `rm -fr ./testfolder` will delete only the folder called testfolder, but if you accidentally insert a space between the . and the `/testfolder`, you will delete everything in your current folder, and attempt to delete the folder called /testfolder!

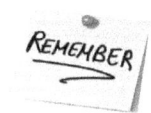

REMEMBER

Every character including "space" at the command prompt means something, and UNIX is case-sensitive.

Every command when typed, needs the ENTER key to instruct the computer to execute the command. Without hitting the ENTER key, the computer waits for more keystrokes to be entered which might constitute your additional parameters or commands.

Let's consider a simple and safe command that does nothing to your files or folders, but merely shows you some information: pwd. It stands for path of working directory (pwd) or some would say, print current working directory to your screen.

So when you have just logged in, the command, pwd (ENTER) will result in the full location of your home directory, the default directory which you will parachute into whenever you login afresh. If you learn later to navigate to another directory, issuing the pwd command will allow you to find out where you are in the directory hierarchy.

Now before we go any further, look at the outcome of issuing the command pwd in the screenshot below. It gives the result /home/eric. We need to introduce to you the concept of a hierarchical folder system in UNIX. All Unix folder systems (including Linux and macOS, which is a kind of UNIX, and Windows operating system, which is just about every single operating system there is to know), use a hierarchical directory structure. For UNIX (and Linux) OS, / indicates the root folder. That is the beginning of everything in the OS. The root folder contains many folders, and we will see in the next section how to view that folder. In this case, our pwd command states that it has the folder called "home". Then there is another "/", called the forward slash. This is followed by "eric". This means that the root folder "/" has a subfolder called "home", which in turn has a subfolder called "eric". To delineate home folder from eric folder, there is a forward slash. There are no spaces in between, and any space is significant. So try not to name your folders with spaces in between. The operating system will have a hard time figuring out whether the space means space between the commands, or space between the name of the folder. Try to use the underscore "_" instead if you need to space out folder names.

The system administrator, gave Eric the username "eric" and assigned his home folder to be, you guessed it, "eric". Hence the home directory when you log in is /home/eric.

Now if you are not Eric, your system administrator will issue you with a username with your name of your choice, most likely. Then your home directory when you enter the pwd command will be

/home/your_user_name

Some people use other types of Linux or UNIXes, and so your home folder may be more complicated. /usr/home/eric or /users/eric or more complicated directory hierarchies. The full hierarchy leading from the root folder "/", is displayed when you enter pwd, which is called the path.

So this is what the p in pwd means, i.e. the meaning of path. So for /usr/home/eric, the path is /usr/home/ and the home folder is "eric" and together the full path directory is displayed when you enter the pwd command.

```
eric — eric@cctv: ~ — ssh eric@cctv.eric.lv — 80×24
[Erics-Mac-mini:~ eric$ ssh eric@cctv.eric.lv
[eric@cctv.eric.lv's password:

The programs included with the Debian GNU/Linux system are free software;
the exact distribution terms for each program are described in the
individual files in /usr/share/doc/*/copyright.

Debian GNU/Linux comes with ABSOLUTELY NO WARRANTY, to the extent
permitted by applicable law.
Last login: Sat Nov 21 09:39:25 2015 from 175.156.9.243
[eric@cctv:~$ pwd
/home/eric
eric@cctv:~$
```

Figure 2.1   Once you type pwd, *and hit the* ENTER *(or* RETURN*) key, the server will display the current operating directory's full path*

In fact, this command is used pretty frequently, especially if your system eventually gets pretty complicated when you create all kinds of folders and sub-folders to store files arising from the many projects and many sequencing runs you might carry out. And you get lost very easily with too many layers of folders. Do not hesitate to immediately enter the **pwd** command to find out where you are. Don't worry, this command does not have any write, modify or delete action and thus it will never affect any of the current folder contents.

```
Command: ls
Action: list out the contents of the current folder
        or directory
```

The **ls** command is not referring to some manga comic but is actually a contraction of the word "**list**". You can list all files within the folder using this command. Enter **ls** and it will return to you a list of all of the objects under your

```
●  ●  ●            🏠 eric — eric@cctv: ~ — ssh eric@cctv.eric.lv — 80×24
[eric@cctv:~$ ls
1000genome    experiment.sh    result
README        hg19.ebwt.1.zip  sratoolkit.2.5.5-ubuntu64.tar.gz
bookExample   refidx           ??????
eric@cctv:~$ ▌
```

Figure 2.2   *Looks like my home directory folder contains too many things. Various attributes of a file, because of the color marking, are immediately obvious. Very striking are the series of question marks or gibberish for some files, and this is because the file names were in Chinese, and the Terminal display could not cope with non-ASCII characters (i.e. anything that is not a–z or A–Z and the usual characters on a typical English keyboard. To avoid this problem, when dealing with NGS datasets, it is recommended that you try to use English names (without spaces please wherever possible). The appendix contains a section on how to deal with non-ASCII characters and spaces in a file or folder name*

current working directory. In some Linux environments, you can program the command line to display certain things such as colours for different types of files and folders, or simply plain white on black, or black on white. So depending on your software configuration established for this connection, there may be various colours of the subfolders (and later we will explain how you might be able to tweak that configuration to change the colours).

Based on my personal configuration, folders are in purple, red is for compressed files, executables are green, and general files which are not specifically labelled turn up white on a black background.

Now this simple `ls` command to display folder contents is missing detailed information. We recommend that one should always be looking out for any signs of trouble by issuing the long form of this command, to display permissions and size of the file and date of its creation. To display the long form of the directory contents, please try entering `ls  -lh` (refer Figure 2.3). (Don't forget the space between `ls` and `-lh`.)

Next, note that the first line shows the total size of all the files in this folder, in K, M or G field, as in kilobytes, megabytes and gigabytes etc. (KB, MB, GB respectively). And then there is the date with time, of when the file was modified, and finally the complete file name. There is something at the beginning with cryptic code like `drw` and hyphens. Well, that is the indication of the file permissions which we will discuss in a later section. Such details are enough for now! These details will come in very handy for giving you a full awareness of what is going on in your folder. So a simple command like `ls`, when modified by parameters such as `-lh`, can become a very powerful command (try `ls  -alh`).

REMEMBER **Commands and its parameters (and/or arguments) are separated by spaces.**

Parameters or flags, are usually prefixed with a hyphen or two: - or --.

Before going any further, you will by now notice that whenever you hit the ENTER key, with or without any commands, something will happen and then you will be returned to what is called the command prompt. Typically, a command prompt will be controlled by the system administrator, as default, and it will display something as plain as a dollar sign "$" or a "#" or a greater-than symbol ">". Sometimes, the system admin is very clever, and will design the command prompt to display a combination of the name of the computer host server you are logged in, or the current date, or the current directory path, or the number of commands you have entered so far! In the Appendix, we have included a section on how you

```
eric — eric@cctv: ~ — ssh eric@cctv.eric.lv — 80×24
[eric@cctv:~$ ls -lh
total 1.7G
drwxr-xr-x 2 eric eric 4.0K Nov 21 09:41 1000genome
-rw-r--r-- 1 eric eric    0 Nov 21 09:44 README
-rw-r--r-- 1 eric eric    8 Nov 21 09:43 experiment.sh
-rw-r--r-- 1 eric eric 1.7G Nov 21 09:55 hg19.ebwt.1.zip
drwxr-xr-x 2 eric eric 4.0K Nov 21 09:42 refidx
drwxrwxrwx 2 eric eric 4.0K Nov 21 09:43 result
-rw-r--r-- 1 eric eric  60M Nov 17 17:03 sratoolkit.2.5.5-ubuntu64.tar.gz
drwxr-xr-x 2 eric eric 4.0K Nov 21 10:00 ??????
eric@cctv:~$
```

Figure 2.3 Note the difference between ls and ls -lh. For each of the folders and files previously listed in ls, it is now one per line

can be as clever as the administrator, and control the command prompt to display whatever you wish (assuming the system allows it), in whatever colour you wish.

Command: cd
Action: change current working directory

You guessed it, cd stands for "change directory", and it does exactly what it stands for. It allows you to switch to another specific folder. To switch to another folder layer, enter "cd ..." where ... is where you key in the name of the folder. In the screenshot, I deliberately use with pwd, so that readers can feel the folder has in fact been switched after the cd command.

REMEMBER At any moment in a UNIX environment, there is a current working directory which all commands will operate in. It is very useful to remember this at the back of your mind whenever you issue UNIX commands: what is the current working directory am I in? pwd will confirm what it is.

Earlier, we suggested you use the command ls -alh

```
eric — eric@cctv: /home — ssh eric@cctv.eric.lv — 80×24
[eric@cctv:~$ cd ..
[eric@cctv:/home$ pwd
/home
eric@cctv:/home$
```

Figure 2.4 *In this case, the* cd *command accepts an argument, which is a subdirectory, or a complete full path. Suppose you entered:*

> cd /home
>
> pwd

It shows /home *as the current working directory. If you enter* ls, *because the current working directory is now* /home, *it will list the folders of only two individuals, because it turns out that this folder contains the home directories of users of this server and currently, there are only two!*

Well, the "a" in the parameter flags (the one prefixed with a hyphen -alh), will display all files, including hidden files. You will notice at the beginning a line with . as a file/directory name, and one with .. next.

If you cd .. what happens when you enter pwd? Where are you now? This .. two points refers to the parent directory as a form of hidden files, it is a shortcut to allow cd to go up one level to the parent folder. So try cd .

Well, "." means current directory and going to the current directory just means you stay put. These shortcuts are really useful so that you don't have to keep keying in the full path.

What other shortcuts are there?

Command	Remarks
cd ..	up one level
cd .	stay put
cd ~	go back to your home directory, in this case cd /home/eric
cd ~/test	jumps to the test folder inside Eric's home folder, i.e. /home/eric/test
cd /	go to the root folder
cd /[TAB]	the TAB key will list out all possible folders in /
cd /h[TAB]	the TAB key will autocomplete "home" if there is only one folder starting with h
cd /h[TAB]e[TAB]	will autocomplete /home/eric/. Wasn't that quick?

The TAB key is your best friend to help you save time and energy.

In addition to folders and files which TAB can autocomplete or show up any ambiguities as a list of options which you can select from, TAB can also autocomplete Linux commands. Try it!

Some folks find keying pwd after cd pretty tedious. So what we have done is to include in the Appendix section a little program you could insert to configure your login prompt so that it will automatically display your current working directory.

Controlling your command prompt

The UNIX login prompt is controlled by the System Variable $PS. The value of this variable will be displayed each time you receive a system prompt. The $PS value is configured in the .bash_login file in your home directory for most UNIXes including Linux which uses the bash shell. The UNIX shell is a long story, but conceptually you can think of it as yet another computer program which is running whenever you log in, your master program so to speak, which takes in whatever you command, and executes it within that shell. Everyone who logs into a server has his own shell. You can create a subshell within your current shell. In fact, every command program you run, runs in a separate shell. Each shell has its own environment and its own set of variables. So the variable $PS controls the prompt display of that shell you are in. It is read from the bash shell start up file. Whenever a new shell starts up, including the login, shell will go into the .bashrc and execute all the commands therein.

By setting PS1 to contain the current working directory, every time you issue a command, and the shell completes the execution and returns a command prompt, it will execute whatever is programmed into PS1.

So if you cd to another folder, and PS1 contains the pwd command, it will automatically display in the prompt, the location of the new current working directory. Without you having to key in pwd — cool!

More advanced explanation:

When Bash starts, it executes the commands in a strict order of different scripts found in the following files.

When Bash is first invoked as an interactive login shell, it first reads and executes commands from the file/etc/profile, if that file exists. After reading that file, it looks for ~/.bash_profile, ~/.bash_login, and ~/.profile, in that order, and reads and executes commands from the first one that exists and is readable.

When a login shell exits, Bash reads and executes commands from the file ~/.bash_logout, if it exists.

When an interactive shell that is not a login shell is started, Bash reads and executes commands from ~/.bashrc, if that file exists. This may be inhibited by using the --norc option. The --rcfile file option will force Bash to read and execute commands from file instead of ~/.bashrc.

This command displays d for the date, and h for the host, and $ as the command prompt.

```
PS1="\d \h $"
```

So that the prompt will look like:

```
Sun Sep 23 host $
PS1="\h \$(pwd) >"
```

So the prompt now looks like

```
host /home/eric >
```

And if you change directory, the prompt will dynamically change as it executes the value of $(pwd) in $PS1 each time a prompt is displayed.

So if you insert the line PS1="\h \$(pwd) >" into .bashrc, whenever you create a non-login shell, you will get the dynamic prompt.

If you insert this line into .bash_profile, then whenever you first login, you will get the dynamic prompt.

You can change colours of your prompt too.

For example,

```
PS1="\e[1;31m\u@\h \e[m \$(pwd)>"
```

This sets the value of PS1 for everything inside the double quotes " ".

\e[starts the colour scheme and \e[m stops it.

1;31m sets the colour to light red. 0;31 dark red, and you guessed it, various numbers other than 31 will give you all kinds of other colours.

\u gives the user name

@ prints the @ symbol

\h gives the host name of the computer.

  (space) prints out a space

$(pwd) executes the command pwd and puts it at that location

> (space) prints out > and a space

For nice tutorials on how to modify your command prompt, google for "How to modify your bash command prompt". A good one may be found here:

http://www.cyberciti.biz/tips/howto-linux-unix-bash-shell-setup-prompt.html

Here are all the colours

http://www.cyberciti.biz/faq/bash-shell-change-the-color-of-my-shell-prompt-under-linux-or-unix/

Remember earlier use of ls shows the various file types thing? If you want to switch to the current directory under a specific folder, enter:

```
cd <folder name>
```

The following command means to change directory to a second folder name which is one level above.

```
cd ../<folder name2>
```

So if there are two folders under home, say eric and john

```
/home/eric
```

```
/home/john
```

From /home/eric if you enter cd ../john, you will essentially be attempting to change directory to John's home directory in /home/john. So this is another use of the shortcut as we discussed earlier. And if John provides you with permission, you can `ls` to view the contents of his folder. If not, there will be an error message, stating

```
ls: cannot open directory : Permission denied
```

I will now discuss a typical scenario when you are doing an NGS project. You really need to keep yourself organized with specific folders for a genome, and a specific folder for each sequencing run, and so on.

Suppose I have a folder 1000genome, and another folder 1001genome and so on until 1011genome. If I want to go to the folder 1010genome, instead of keying all 10 characters, I will just enter `cd 1` and then press TAB, and I will see all 12 folder names and TAB would have autocompleted only one possibility, and that is 10 because there is no other possibility. So it allows me to continue typing another 1, and if I press TAB again, `cd 101[TAB]` it will return only two possibilities, 1010genome and 1011genome. So pressing 0 and then TAB, `cd 1010[TAB]` will autocomplete genome and pressing enter, I will go straight to the folder 1010genome. So instead of typing all ten characters and make mistakes, I only needed to key in `cd 1[TAB]11[TAB]`, just five keys, and I can autocomplete the correct folder. This may seem like a trivial example, but imagine if there are folders with really long names. Typing them over and over again can be extremely tedious.

So this technique is very useful and can reduce the input of human error, especially with very long file names.

```
Command: mkdir
Action: make directory
```

If you have too many files and too complex a folder to lump them all into one flat structure, what you need is the ability to create new directories, each with intelligible names. Use the `mkdir` command, which takes in directory names as argument.

```
mkdir <foldername>
```

So if you wish to create a folder called rawData, `mkdir rawData` is the command to use.

Note carefully about UNIX case-sensitivity we told you about at the beginning of this chapter. rawData, rawdata and Rawdata are three completely different folders, so please be careful.

To illustrate this command, take a look at the following screenshot.

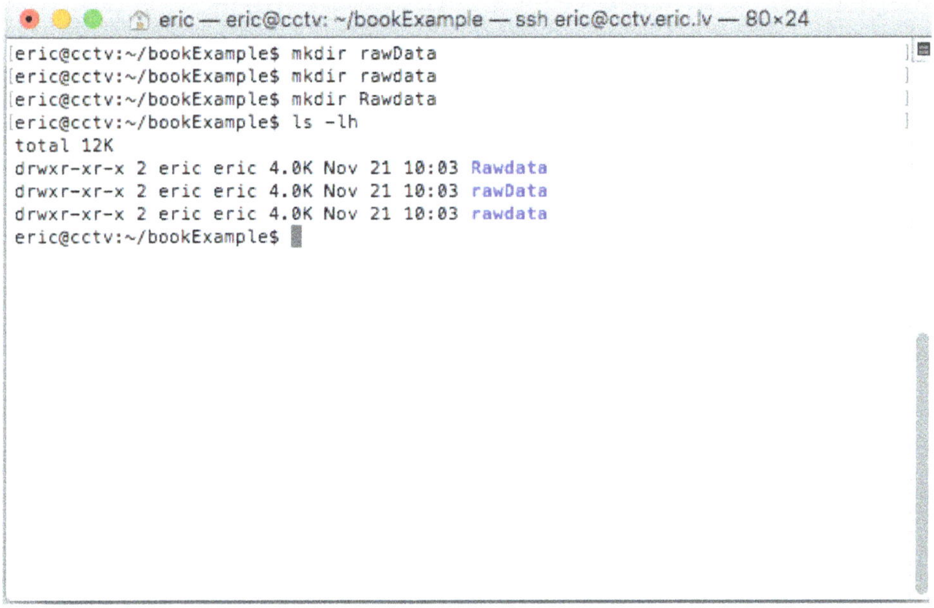

```
●  ●  ●      eric — eric@cctv: ~/bookExample — ssh eric@cctv.eric.lv — 80×24
[eric@cctv:~/bookExample$ mkdir rawData
[eric@cctv:~/bookExample$ mkdir rawdata
[eric@cctv:~/bookExample$ mkdir Rawdata
[eric@cctv:~/bookExample$ ls -lh
 total 12K
 drwxr-xr-x 2 eric eric 4.0K Nov 21 10:03 Rawdata
 drwxr-xr-x 2 eric eric 4.0K Nov 21 10:03 rawData
 drwxr-xr-x 2 eric eric 4.0K Nov 21 10:03 rawdata
 eric@cctv:~/bookExample$ ▓
```

Figure 2.5   *Creating three different directories rawData, rawdata and Rawdata, and using* ls *to show that they have been successfully created as three independent folders*

Note that you could also have issued a single command:

```
mkdir rawData rawdata Rawdata
```

mkdir takes in multiple valid folder names as arguments and treats them as if you wanted to create all of them in one go. So please be careful, in case you have created too many folders which you don't really need.

```
Command: rmdir
Action: delete or Remove Directory
```

If we can create something, we must be able to remove it. Let me demonstrate how we can delete Rawdata and rawdata. Unused directories can create confusion. To delete folders, please use the rmdir command as follows:

```
rmdir <foldername>
```

Use the following commands:

```
rmdir Rawdata

rmdir rawdata

ls
```

You can see that only rawData is now left.

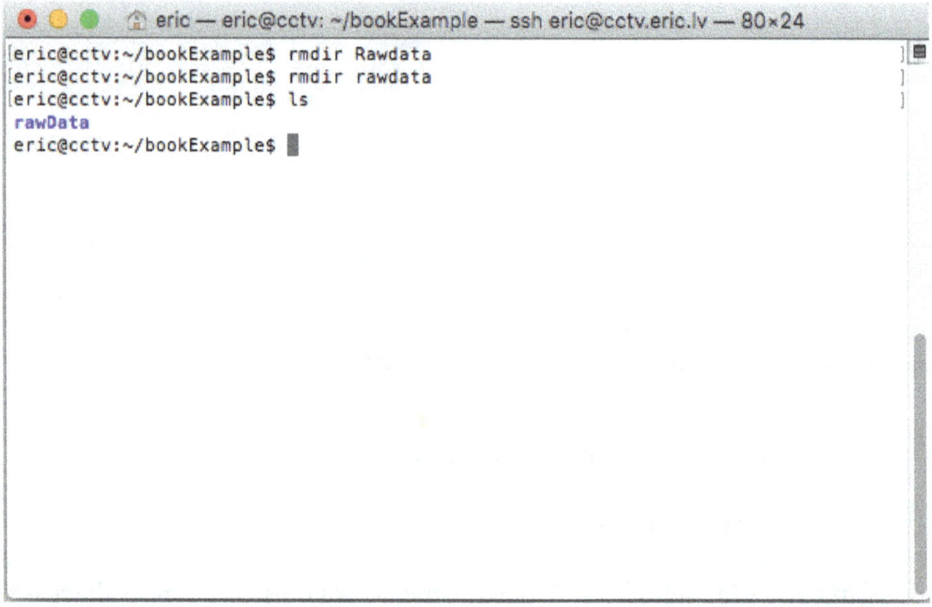

Figure 2.6 `rmdir Rawdata` *and then* `rawData`, *followed by* `ls` *to confirm that only rawData is left*

There is yet another shortcut which you really need to know. Press the arrow up key. Miraculously, you can see `ls` appear. Press the arrow up key again and again, and you will see all previous commands which you have just keyed, all in sequence and in the same order as you have entered the commands. If you can scroll back, you can also scroll forward, and you have guessed it. Press the arrow down key to scroll down. So pressing arrow up and arrow down, you can essentially save time and select exactly which previous commands you wish to repeat.

For example, let us scroll up to the `mkdir` commands, and recreate the rawdata and Rawdata folders. It only takes a few keystrokes of the arrow up or down key to find them, and hit ENTER to execute them. `ls` to check that you have now created all three folders.

Now, let us try another interesting variation of the `rmdir` command

```
rmdir rawdata rawData Rawdata
```

You will see that all three folder names can be used as arguments for `rmdir`. In other words, `rmdir` can take more than one argument, just like `mkdir` as discussed above. So in one single command, all three folders can be deleted.

Let us recreate them again. But it is really tedious to rekey in all these words. So, there is yet another shortcut. Press the arrow key up, to bring back the `rmdir rawdata rawData Rawdata` command.

Note that `rmdir` and `mkdir` differ only in the r and the k. So, if you use the left and right arrow key, you can actually navigate to the position where you want to make the change, and BACKSPACE to remove the r character, and insert the k, thereby creating the command:

```
mkdir rawdata rawData Rawdata
```

Without having to type everything. Try it! Amazing isn't it?

Now, if you feel that having to press so many arrow left keys is too tedious, there is yet another shortcut, the `Control-A` command. `Control-A` or `Ctrl-A` means to press the CONTROL key, and without letting go, press the A key.

Now that you have created all three, let us delete all three folders again. Arrow up once to recall

```
mkdir rawdata rawData Rawdata
```

`Ctrl-A` to jump to the beginning of the line.

```
rmdir rawdata rawData Rawdata
```

The cursor will jump to r. Arrow key right, twice and insert k. `Ctrl-A` again, and arrow key right once and press DELETE or BACKSPACE to delete the r. In this way, the following command can be easily edited.

```
mkdir rawdata rawData Rawdata
```

Hit ENTER and voila, all three folders will get created all in one go. To delete them, arrow key up twice and find the `rmdir` command and hit ENTER. So you can see you can do and redo or undo commands just by editing them on the command line. Pretty cool yes?

There are even more time saving shortcuts, and more powerful Linux commands, particularly those using the bash shell. We will discuss these in the Appendix in greater technical detail, and later on.

As a teaser, how would you create a series of folders with names sequentially from 1000genome to 1234genome, i.e. 1000genome, 1001genome, 1002genome, etc.

Hint: `mkdir {1000..1234}genome` and to remove all of them?

`rmdir {1000..1234}genome` and you will notice all of them were removed if you check with `ls`. How about `mkdir 1{000.234}genome`?

That does the trick too! (Note this trick is called "brace expansion".) We will discuss the wildcard * and ? in the next section.

2.3 File Operations

I hope the previous section about folders and their manipulation did not make you dizzy. In this section, we are just going to describe three little file operations: copy, delete and move.

```
Command: cp
Action: copying files
```

To copy files, the command is `cp`, and its simple operation is:

```
cp <oldfile> <newfile>
```

Suppose you have a sequence file whose original name is seq1.fastq, and you wish to keep it pristine and make a working copy seq2.fastq for various manipulations.

```
cp seq1.fastq seq2.fastq
```

In order to demonstrate the commonly expected situation, in the example below, I deliberately named a new file name with exactly the same name as the original, resulting in an error message. A folder cannot contain two files with exactly the same name. So remember to copy the files you want to duplicate with a new file name. I know many people will experience this problem because in the previous section, I taught you how to press **TAB** to autocomplete a bash command, and the result is that it is all too easy not to enter two different file names.

Just to share with you my experience and working habit, I typically copy a file like seq1.fastq to a file like seq1.fastq.bak, so that I can still use the **TAB** auto

```
● ◎ ● ⌂ eric — eric@cctv: ~/bookExample/rawData — ssh eric@cctv.eric.lv — 80×24
eric@cctv:~/bookExample/rawData$ ls
seq1.fastq
eric@cctv:~/bookExample/rawData$ cp seq1.fastq seq1.fastq
cp: 'seq1.fastq' and 'seq1.fastq' are the same file
eric@cctv:~/bookExample/rawData$ cp seq1.fastq seq2.fastq
eric@cctv:~/bookExample/rawData$ ls
seq1.fastq  seq2.fastq
eric@cctv:~/bookExample/rawData$ ▊
```

Figure 2.7   *First, we enter* ls *to list out all files in the folder. To create a file, we could use the UNIX command* "touch seq1.fasta" *with nothing inside of course as this is for demonstration purposes. With only one seq1.fastq, then* cp seq1.fastq seq1.fastq *will trigger an error message, in that we cannot copy the contents of a file into another file with the same name*

completion and quickly add in the .bak extension, bak standing for backup. Having a consistent extension can be very helpful as we shall see later.

I know everyone liked the shortcuts we described, what we can also call keyboard cheats. So for the above error, the executed instruction which triggered an error can be recalled by an up arrow keystroke. Then you can navigate by the left arrow key to the position of the second **1** and replace it with **2**. As many times as you wish, you can recall old commands from the so-called history of your commands stored in a hidden file called ".bash_history" and jump to the front of a few multi-executed commands and make the necessary modifications and re-execute the desired command.

> Command: rm
> Action: delete or Remove Files

Besides "delete" in English, "remove" may seem like a gentler command, but their effect in UNIX is the same — the file removed or deleted is gone forever. So do be careful when using the rm command, or for that matter, the previously described rmdir command. So if, for example, I now feel seq2.fastq takes up a lot of space (real sequence data are big, as later chapters will show), for housekeeping purposes, enter the following to delete it:

```
rm seq2.fastq
```

Henceforth the file seq2.fastq would have disappeared from your hard drive forever. Note that rm coupled together with the -r parameter, can be used to remove a folder. If the folder you want to remove still contains files or other folders, for rmdir, an error message appears, indicating that the folder you want to remove is not empty. To remove say a temporary folder called temp containing files or other folders inside, enter rmdir temp because rmdir will simply trigger an error — rmdir: temp: Directory not empty. In fact, rm temp without the -r parameter will also trigger an error if you are trying to delete a folder.

```
rm: cannot remove 'temp': Is a directory
```

Only rm -r will smoothly remove temp and everything inside.

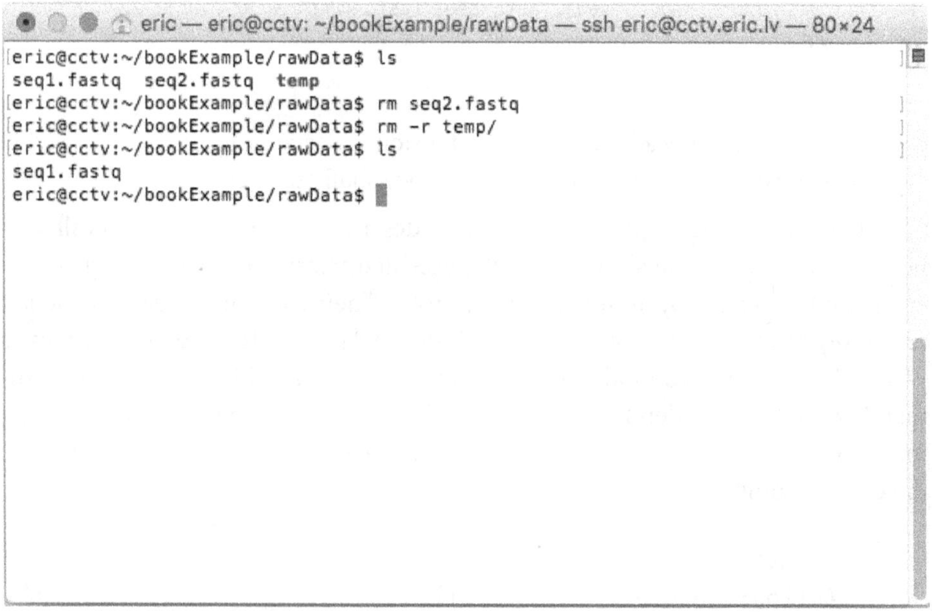

```
● ○ ●  ⓘ eric — eric@cctv: ~/bookExample/rawData — ssh eric@cctv.eric.lv — 80×24
[eric@cctv:~/bookExample/rawData$ ls
 seq1.fastq  seq2.fastq  temp
[eric@cctv:~/bookExample/rawData$ rm seq2.fastq
[eric@cctv:~/bookExample/rawData$ rm -r temp/
[eric@cctv:~/bookExample/rawData$ ls
 seq1.fastq
 eric@cctv:~/bookExample/rawData$ ▮
```

Figure 2.8 *The folder rawData contains seq1.fasta and seq2.fasta, and another folder temp.* rm -r seq2.fasta temp *will sequentially remove all files and folders leaving only seq1.fasta*

```
Command: mv
Action: move or rename a file
```

The instruction to move a file is `mv`. It should not be hard to imagine its purpose is to move the file to another folder. Not surprisingly, the command is: `mv <file name> <folder name>`. For example, to move this file seq1.fastq to a pre-existing folder called "original" is:

```
mv seq1.fastq original/
```

If the folder does not exist, the bash shell will complain.

```
mv seq1.fasta original/
mv: cannot move 'seq1.fasta' to 'original/': Not a
directory
```

You will need to create the folder first

```
mkdir original mv seq.fasta original/
```

The `/` after `original` refers to original as a folder. Suppose you have not created the folder "original" yet, and you entered the command without original with the `/` (no spaces in between).

```
mv seq1.fasta original
```

This command will succeed. What has happened here? `mv` would have renamed seq1.fasta into another file called "original". In other words, if there is a name of a pre-existing folder and you `mv` a file to that folder name, `mv` will guess that you want to move that file into the folder, even though you did not enter the `/` behind.

So from this simple exercise, we have discovered the rename function for UNIX. There is no need for another command called "rename" just for renaming file1 to file2. `mv file1 file2` suffices.

The following shows how the raw data to the renamed folder, inside seq1.fastq renamed seq1.fastq.bak. Command is:

```
● ● ●  ⌂ eric — eric@cctv: ~/bookExample/rawData/raw — ssh eric@cctv.eric.lv — 80×24
[eric@cctv:~/bookExample/rawData$ ls
original  seq1.fastq
[eric@cctv:~/bookExample/rawData$ mv seq1.fastq original/
[eric@cctv:~/bookExample/rawData$ ls
original
[eric@cctv:~/bookExample/rawData$ ls original/
seq1.fastq
[eric@cctv:~/bookExample/rawData$ mv original/ raw/
[eric@cctv:~/bookExample/rawData$ ls
raw
[eric@cctv:~/bookExample/rawData$ cd raw/
[eric@cctv:~/bookExample/rawData/raw$ ls
seq1.fastq
[eric@cctv:~/bookExample/rawData/raw$ mv seq1.fastq seq1.fastq.bak
[eric@cctv:~/bookExample/rawData/raw$ ls
seq1.fastq.bak
eric@cctv:~/bookExample/rawData/raw$ ▊
```

Figure 2.9

1. First, use the ls command to verify that the current working folder contains a folder "original" and seq1.fastq, the file.

2. mv seq1.fasta original

3. Once again ls, to see if the move was successful, i.e. the seq1.fasta would have disappeared from the current directory.

4. ls original, which lists the contents of the folder, "original". (ls also can be used directly for specific sub-folders display all contents)

5. mv original raw will rename original into raw. (so far bookExample/ rawData/ are still in this folder). Using the cd command to switch to the folder "raw", and then use ls to list all items therein.

6. Finally, the unique file seq1.fastq is renamed seq1.fastq.bak using mv command.

The mv command also applies to folders. For example, to change the folder called "original" folder into a folder called "raw", that is "mv original/ raw/", it does not matter if there are files or folders, mv will move them accordingly.

Just a note of caution. If you typed too quickly and accidentally inserted a space between original and /, mv will attempt to do as you instructed. It will try to move the folder "original" to the root folder "/", which will trigger an error, i.e. you do not have permission to write into that folder. mv: cannot move 'original' to '/original': Permission denied. The next section will talk about this thing called "Permissions".

But before that, I will introduce a series of powerful shortcuts. Suppose you have to create three folders genome1001. genome1002 and genome1003 for each of your genome projects. You could very quickly use a combination of typing and the shortcuts you were introduced to earlier.

```
mkdir genome1001

[arrow up] [backspace] 2 [Enter]

[arrow up] [backspace] 3 [Enter]
```

But what if you have a hundred folders you need to create subsequently, from genome1003 to genome1103?

There is a more powerful shortcut called brace expansion mentioned briefly before.

```
mkdir genome{1004..1103}

ls -l
```

You will find 103 folders named as you expected. That sure beats having to manually create 100 more folders, even using the history and editing shortcuts.

Suppose you made a mistake. You want to rename genome1001 into g1001. You could `mv genome1001 g1001` which would work. But what if the rest of the folders are also wrongly named. At the beginning, when nothing was copied into the folders you could

```
rm -r genome*
```

And recreate the folders afresh

```
mkdir g{1002..1103}
```

But if you discovered this too late and files have already been deposited into these other folders? How would you execute

```
mv genome1001 g1001

mv genome1002 g1002
```

etc., a hundred or more times?

Now some of you may be tempted to use a wildcard character *

```
mv genome* g*
```

Unfortunately it does not work as you think. When the bash shell encounters an asterisk * or a wildcard character, it performs filename expansion, very similar to the TAB auto-complete command, which shows up a whole bunch of options. Here, before mv is executed, bash shell expands the command from mv genome* g* to mv genome1001 genome1002, etc., till genome1103 and again for g* genome1001 all over to genome1103! Note that the second set of folders are expanded because they also match g* wildcard.

This command now has 206 arguments! Then shell proceeds to attempt to move all the folders to the last folder genome1103, as instructed bt shell after the filename expansion. So the first genome1001 to genome1102 will be successfully moved into the last folder of the second set, genome1103, but when mv attempts to move genome1103 to itself (the last folder in the expanded command), it triggers an error. Then it attempts again to move genome1001 to genome1102 to genome1103, but they have already been moved, so shell will complain "No such file or directory" for another 102 times!

To reverse this flawed operation, you could use the * correctly as follows.

```
mv genome1103/* .
```

This expanded will mean mv genome1103/genome1001 genome1103/genome1102 ... . (Recall that . refers to the current directory.) So essentially this command with 1102 arguments plus . the last argument, will essentially move everything back to where we started, the current folder with 103 folder.

So beware of the power of the wildcard *. Always think in terms of wildcard filename expansion at the command line. There is of course a limit to how many arguments a command line command can take. If this limit is exceeded, shell will complain — Too many arguments.

So how do we solve this problem? In one line!

```
for x in genome*; do mv $x ${x/genome/g}; done
```

This is actually a bash program all squeezed into a single line. We will discuss simple bash scripting in the Appendix, as it will be frequently needed in managing NGS projects. Knowing some bash scripting will go a long way towards high throughput work!

We will discuss filename expansion in greater detail later on, but certainly, you can see how powerful the 1001..1103 command is, the double dot .. in this context of being sandwiched between two numbers, also known as the range operator, will essentially instruct bash to expand the arguments from 1001 to 1103.

2.4 Assignment of Permissions

The Linux operating system is known for its strong security design. One reason is that it has a comprehensive permissions or rights management system. If you do not have access rights or do not have permission to execute a command, especially manipulating the operating system core files themselves, you simply cannot do anything that can damage the operating system. Your scope is merely confined to the folders and sub-folders you have access rights to. So if a virus or malware is executed, it is certainly because the administrator privileges or "root" has been stolen. Here we share two commands about permissions, the file permissions changed to a different owner and changed to a different mode, the execute permissions for the file.

```
Command: chown
Action: change ownership to another user
```

For the command chown, own refers to owner and "ch" refers to change. We can therefore deduce that chown means to change the file permissions to another. So it is not surprising that it takes two arguments

```
chown newowner file_or_foldername
```

In the execution of this command, sometimes the system does not allow you to proceed. One way to get around this problem is to use sudo privileges to operate this command. The system administrator determines whether users have sudo privileges and they can add you to the administrator's list of sudoers. sudo means "superuser do" and sudoers means the users who can do what a superuser can do. Before I explain privileges, let me demonstrate.

My example which is to make the owner of seq1.fastq.bak change from eric to root.

The first time the command is executed, chown root seq1.fastq.bak the system rejects it.

```
chown:  changing  ownership  of  'seq1.fastq.bak':
Operation not permitted
```

If we now add sudo in front of the command sudo chown root seq1.fastq. bak the bash shell will ask for your password, and if you are in the sudoers' list, the command will be completed successfully.

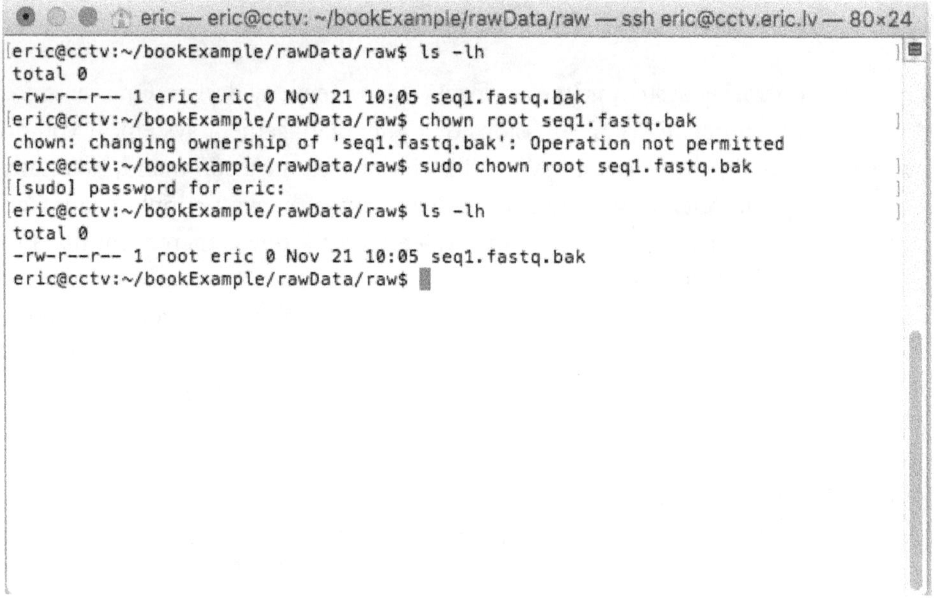

```
● ○ ●  ⌂ eric — eric@cctv: ~/bookExample/rawData/raw — ssh eric@cctv.eric.lv — 80×24
[eric@cctv:~/bookExample/rawData/raw$ ls -lh                               ]▤
 total 0
-rw-r--r-- 1 eric eric 0 Nov 21 10:05 seq1.fastq.bak
[eric@cctv:~/bookExample/rawData/raw$ chown root seq1.fastq.bak            ]
 chown: changing ownership of 'seq1.fastq.bak': Operation not permitted
[eric@cctv:~/bookExample/rawData/raw$ sudo chown root seq1.fastq.bak       ]
[[sudo] password for eric:                                                 ]
[eric@cctv:~/bookExample/rawData/raw$ ls -lh                               ]
 total 0
-rw-r--r-- 1 root eric 0 Nov 21 10:05 seq1.fastq.bak
 eric@cctv:~/bookExample/rawData/raw$ ▮
```

Figure 2.10   *1. Originally, the owner of seq1.fastq.bak is eric. We want to change permission to root. This was rejected.*

2. After prepending sudo to the command, the change in ownership was successfully executed.

3. Once again ls *to view the file details, and you will find the owner has become root.*

You may be confused by now what this is all about. To get a better understanding of ownership and privileges, let us go back to the ls -alh command.

Explanation of the directory listing, levels of user rights, group and permissions.

When you perform the long form of the directory listing, ls -l, you will see the full description of each file or folder in your current working directory. This will have a few seemingly strange letters at the beginning of the line, like the following:

```
ttw:/home/tinwee > ls -alh
total 458K
drwxr-xr-x 9    tinwee    users    1.0K  Sep 29    14:48    ./
drwxr-xr-x 11   root      root     1.0K  Sep 10    15:44    ../
-rw-r--r-- 1    tinwee    users       0  Sep 15    03:17    .addressbook
-rw------- 1    tinwee    users     39K  Sep 29    22:57    .bash_history
-rw-r--r-- 1    tinwee    users      98  Sep 29    14:03    .bash_profile
-rw-r--r-- 1    tinwee    users     126  Sep 29    14:04    .bashrc
drwx------ 2    tinwee    users    1.0K  May 21    17:02    .screen/
drwx------ 2    tinwee    users    1.0K  May 10    21:26    .ssh/
drwxr-xr-x 2    tinwee    users    1.0K  May 28    01:40    conv/
drwxr-xr-x 2    tinwee    users    1.0K  May 20    22:58    ex/
```

The first letter, d, means directory. So conv, ex are subdirectories of my home directory /home/tinwee. .ssh and .screen are hidden subdirectories, and . refers to the current directory and .. refers to the directory one level up, i.e. /home which obviously is owned by root or the root group, whereas all the files and folders that belong to me are owned by me tinwee of the group users.

The next three blocks of three characters each, refer to the permissions status of user, group and others consecutively. r stands for read, w stands for write, and x stands for executable. For a directory to be visible, it has to have the x permissions set. If it is set to a -, then it is denied. Therefore, current directory, the home directory /home/tinwee, has drwxr-xr-x, actually means that it is a directory, and has permissions for user to rwx read-write-execute, i.e., I can perform everything with the full permissions. It also means that r-x for both group and others, such that all users belonging to the group "users" will be able to read and execute this directory, but not write. The same goes for all others.

For .screen and .ssh, my private parameter files for my logins, obviously, I do not want anyone snooping into these folders. So the permissions are set at drwx------ and the dashes denied all rights to the group and others. No one else besides myself can view the contents or browse these two directories.

Similarly for files, such as the history of all my commands I have executed on this machine, -rw------- means that it is not a directory, and it has read-write permissions for myself as a user. It is not an executable program, but is just a text file recording the history of my commands executed at the command line prompt. And everything else is denied. Nobody else can read, write or execute it.

To change these permissions, we have to use the chmod command.

```
Command: chmod
Action: change mode of the file/folder permissions
```

As a bioinformatics user, you would frequently have the opportunity to download from the Internet bioinformatics tools and applications. After the download, you may not necessarily have the execute permissions for the program file. So when you enter the command to execute it, the file permissions will block the bash shell from executing it. To make it execute, we have to change the file permissions to executable, i.e. change its mode to x for the user.

```
chmod u+x <file name>
```

So a file with permissions -rwr--r-- will become -rwxr--r--. If we wish to change the mode to executable for everyone, we can simply run

```
chmod +x <filename>
```

chmod means change mode and the +x parameter refers to confer execute permissions for user, group and others. So a file with permissions -rwr--r-- will become -rwxr-xr-x.

For example, a research process script which you might be able to write (a script on a research design process will be explained in the following sections), whose file name is runWorkflow.sh, needs to be executed after downloading. The original downloaded file will have its default permissions set as -rw-r--r-- (this is called the umask). This means that the permissions status of the downloaded program file is that it is simply a text file. Now to make it executable, we need to change its permissions to executable. The command is:

```
chmod +x runWorkflow.sh
```

For most users, after first contact with the permission mode, in addition to execute permissions, there will be other read-write needs, so just understanding +x is not enough. For example, if we are totally paranoid, we should change our file permissions for all really important files with no permissions at all. But typically, UNIX by default, sets up a directory with 777 privileges, i.e. rwxrwxrwx, and for files, 666 i.e. rw-rw-rw-.

```
chmod 777 filename
ls -l filename
-rwxrwxrwx .... filename
```

Now this is a geek's way of making things computer processable but totally opaque to beginners. chmod 777 means full permissions at all for user, group and others respectively for each digit.

```
eric — eric@cctv: ~/bookExample — ssh eric@cctv.eric.lv — 80×24
eric@cctv:~/bookExample$ ls
rawData   runWorkflow.sh
eric@cctv:~/bookExample$ chmod +x runWorkflow.sh
eric@cctv:~/bookExample$ ls
rawData   runWorkflow.sh
eric@cctv:~/bookExample$
```

Figure 2.11   *Change the executable mode, using* chmod +x. *To revert it back to a non-executable mode,* chmod -x runWorkflow.sh *So the plus or minus sign actually allows you to switch on or off the executable mode*

This is the octal way of saying symbolically

chmod ugo+rwx <filename>

i.e. user, group and others, allow full read, write or execute permissions. Here is a table showing what the octal numbers mean.

Table 2.1   *Octal number of permission Representation*

Octal	Value permission
7	read write and execute
6	read and write
5	read and execute
4	read only
3	write and execute
2	write only
1	execute only
0	no permissions

I personally don't bother with lumbering my brain with these numbers. I just use the symbolic values, namely ugo+ or -rwx in the setting of my modes because I know exactly what I want to set. But unfortunately, by default, most Linuxes set the umask in number format, umask 022 for root and for users (sometimes 002).

Try the command

```
grep umask /etc/profile
```

grep will pull out the lines containing umask, and indeed, your Linux is probably configured with umask 022.

Now this is where things get a bit confusing. umask means when you set up a file with default 666, subtract 022 from it, to give you 644. Then set the permissions of the newly created file as 644, equivalent to manually keying:

```
chmod 644 newfile

ls -l newfile

-rw-r--r-- ... newfile
```

This means everyone can read the file, but only the user, yourself, can write into it. For directories, the default is 777 i.e. drwxrwxrwx full permissions. But with an umask of 022, 777 − 022 = 755, which means drwxr-xr-x.

One way of remembering the table, if you have to punish yourself in this way: execute permissions is worth 4, write permissions is 2 and read permissions is 1. So

```
rwx = 4 + 2 + 1 = 7.

rw- = 4 = 2 = 6.

r-x = 4 + 1 = 5.

r-- = 4
```

Each digit represents a different identity, such as user file owner (u), group (g) and others (o) in this order. 777 is rwxrwxrwx, i.e. all three categories of user identity can get full access. If you want to set only the user file owner alone to obtain full read-write access but group users and others only read permissions, you should set mode to 644 (4 + 2,4,4) i.e. rw-r--r--, which is standard for most configurations as the default when you create a new file that is not executable. Similarly 755 (4 + 2 + 1, 4 + 1, 4 + 1) will be equivalent to rwxr-xr-x, which means the file will be executable, in addition to being readable by all, and only writable to the user as its owner.

So what is the practical usage of this? Remember the command which I showed you earlier as a bash one-liner program. For x in `genome*` ; do `mv $x $(echo $x | cut --complement -c2-6)`; done. Well, you can save this command into a file, say createfolders.sh and `chmod +x createfolders.sh`.

This command can now be executed repeatedly.

```
./createfolders.sh
```

And it will change the folders from genome1001 to g1001, etc. So if you have a series of commands which you frequently use in the same sequence, you can save them into a file, make that file executable, and you can execute these series of commands in a single command line by invoking that file name just like a UNIX command.

And if you still think that is too much typing, well, you could abbreviate it further using the alias command.

```
alias crf='/home/tinwee/createfolders.sh'
```

And henceforth, in this session of Linux login at least, entering `crf` is the same as running the command `/home/tinwee/createfolders.sh`

```
Command: alias
Action:  create  an  alternative  way  of  naming  a
         command.  Alternatively,  you  could  use  a
         softlink instead

Command: ln
Action:  softlinks  a  file  with  typically  a  shorter
         name
```

```
ln -s /home/tinwee/createfolders crf
```

This will softlink `crf` to `createfolders`. Now if there was another command somewhere that was called `crf` too, what will happen now? Softlink method: it only refers to the `crf` and the `createfolders` in `/home/tinwee`.

The path

Suppose there is another command crf in another folder, say in /usr/local/bin/crf, which crf will be used will depend on the order of the directories listed in the environmental variable $PATH.

```
echo $PATH
```

And see the sequence which shell searches for executables. Usually, for commands and programs in your current directory, we will include the dot . in the PATH as the very last item in the queue. So the first directory will be searched for executable programs and so on, until the final directory, which is the current working directory, as represented by the dot.

So if /usr/local/bin folder precedes /home/tinwee as represented by the dot, then the crf program in there will take precedence when bash shell decides which crf to execute.

How to add a new path?

Suppose you have a new program which you have installed, and you have kept it in a folder called /home/tinwee/bin/. Simply use the command:

```
export PATH=$PATH:/path/to/dir1:/path/to/dir2:/home/tinwee/bin:.
```

This takes whatever pre-existing in the variable $PATH and adds dir1, dir2 and /home/tinwee/bin:. to the search path whenever shell is looking for executables. If you do not place the . somewhere, then programs which you compile in your current directory, you have to explicitly state the full path, or use the following idiom: Add dot slash "./" in front of the new program name to signify that it is this full path and this program which you wish to execute, to prevent bash shell from traversing the directory path only not to find any program called "newprogram" and trigger an error "program not found"

```
./newprogram
```

And one more thing, instead of having to key in this export command all the time to amend the PATH, you should append this command into a hidden file called .bashrc (and that is with the dot in front of the bashrc) such that every time a shell is created, the path is changed to include your new folders containing executables.

2.5 Understanding System Status

Bioinformatics is frequently a computationally resource-intensive business. For those unfamiliar with computer hardware resource limitations, once you encounter a problem where your computer cannot load the program, takes too long, or runs out of memory, you are unlikely to be able to figure a way out. To avoid the unnecessary waste of time, when a server crashes, or swaps with hard disk excessively, you can ask the administrator and wait for their trouble ticket to resolve your problem. Alternatively, wouldn't it be better if you knew something about such hardware limitations and have a better understanding through learning about the specifications of your computer? (Some administrators actually restrict users from knowing the full details of your computer server you are logged on, and block the programs which queries system status.) By comparing the specifications of bioinformatics tools and their hardware requirements through reading their system requirements documentation, we can best avoid the futile or the inevitable.

```
Command: cat /proc/meminfo
Action: print out Memory Information
```

The specifications of the computer's physical memory is stored inside a file called /proc/meminfo and you can read its contents using this command:

```
cat /proc/meminfo
```

Strictly speaking, this is the system memory information that is being displayed. The `cat` command which displays the contents of the `meminfo` file to the terminal display will be explained later. Look at the first line of the output. For example, my machine is 16,439,936KB, after conversion 16,439,936 by 1024, the answer is 16054.625MB; once again dividing by 1024, it rounds up to about 15.7 GB. That is, this machine is specified at 16 GB of physical memory by the computer manufacturers.

Print memory information. It is important how much physical memory is available to the end-user of any computer, and the key is the first line `MemTotal`. Then look at the second line, `MemFree`, which tells you how much idle memory is left.

```
Command: cat /proc/cpuinfo
Action: print processor information
```

```
● ○ ●              ⌂ eric — root@galaxy: ~ — ssh root@10.12.2.157 — 80×24
[root@galaxy:~# cat /proc/meminfo                                              ]▤
MemTotal:        16439936 kB
MemFree:          2442748 kB
Buffers:            52896 kB
Cached:          13380336 kB
SwapCached:             0 kB
Active:           9097560 kB
Inactive:         4418940 kB
Active(anon):       74428 kB
Inactive(anon):      9628 kB
Active(file):     9023132 kB
Inactive(file):   4409312 kB
Unevictable:            0 kB
Mlocked:                0 kB
SwapTotal:        3906556 kB
SwapFree:         3906556 kB
Dirty:                  8 kB
Writeback:              0 kB
AnonPages:          82836 kB
Mapped:             39480 kB
Shmem:                788 kB
Slab:              428772 kB
SReclaimable:      415704 kB
SUnreclaim:         13068 kB
```

Figure 2.12   *Print installed memory information by "cat" a system file*

The era of the multi-core processor has arrived. Regardless of whether the server processor inside is working with real or virtual multi-core, you can check out from inside the Linux OS. But as for servers, if we do not have the decision-making ability to deal with hardware purchases and upgrades, we will not consider pursuing the issue of relative benefits of the virtual with the real. To query the processor specifications, enter:

```
cat /proc/cpuinfo
```

Once the output completes with a few dozen lines on the screen, scroll to go back and look at several key things. Identify the processor and model name.

For example, my computer's processor number is 7, meaning that this computer has eight cores, because Linux counts from 0: 0,1,2 ..., 6 to 7. The model name in my case is Intel (R) Xeon (R) CPU E5-2630L v2 @2.40GHz, so it is clear that its Intel's Xeon processor — an Internet search for this model will give you detailed information of the original specifications as well. Of course! This command cannot be perfect, so specifications such as processor, operating system are too new, the program may not necessarily detect the correct model, so beware!

```
eric — eric@cctv: ~/bookExample — ssh eric@cctv.eric.lv — 80×24
eric@cctv:~/bookExample$ cat /proc/cpuinfo
processor       : 0
vendor_id       : GenuineIntel
cpu family      : 6
model           : 62
model name      : Intel(R) Xeon(R) CPU E5-2630L v2 @ 2.40GHz
stepping        : 4
microcode       : 0x1
cpu MHz         : 2399.998
cache size      : 15360 KB
physical id     : 0
siblings        : 1
core id         : 0
cpu cores       : 1
apicid          : 0
initial apicid  : 0
fpu             : yes
fpu_exception   : yes
cpuid level     : 13
wp              : yes
flags           : fpu vme de pse tsc msr pae mce cx8 apic sep mtrr pge mca cmov
pat pse36 clflush mmx fxsr sse sse2 ss syscall nx pdpe1gb rdtscp lm constant_tsc
 arch_perfmon rep_good nopl eagerfpu pni pclmulqdq vmx ssse3 cx16 pcid sse4_1 ss
e4_2 x2apic popcnt tsc_deadline_timer aes xsave avx f16c rdrand hypervisor lahf_
```

Figure 2.13   *To find out about your PC's hardware information, you do not need to dismantle the machine casing, the OS would have extracted this information. Just enter* cat /proc/cpuinfo

> Command: df
> Action: disk space information

Disk space in biocomputing is also very important, not just because you need a lot of storage for original data. The computation process often requires sufficient disk space for temporary storage, or when memory is running low, to use disk space as a memory buffer for what Linux calls the swap mechanism. The operator of bioinformatics tools really needs to understand the amount of disk space needed. Generally, if a particular computation takes a long time, users tend to enter the command into a terminal and thereafter leave the computer to let it run on its own, thinking that coming back a few hours later one can collect the output data. Unfortunately, they do not realize that a disk-intensive calculation may be halted or crash due to insufficient disk space. Your program may have long been interrupted, and so, do make your calculations prior to running a very big computation job.

```
● ● ●       🏠 eric — eric@cctv: ~/bookExample — ssh eric@cctv.eric.lv — 80×24
[eric@cctv:~/bookExample$ df
Filesystem      1K-blocks       Used Available Use% Mounted on
/dev/vda1       20504916 17556644    1892576  91% /
udev               10240        0      10240   0% /dev
tmpfs             101268    12584      88684  13% /run
tmpfs             253168        0     253168   0% /dev/shm
tmpfs               5120        0       5120   0% /run/lock
tmpfs             253168        0     253168   0% /sys/fs/cgroup
[eric@cctv:~/bookExample$ df -h
Filesystem      Size  Used Avail Use% Mounted on
/dev/vda1        20G   17G  1.9G  91% /
udev             10M     0   10M   0% /dev
tmpfs            99M   13M   87M  13% /run
tmpfs           248M     0  248M   0% /dev/shm
tmpfs           5.0M     0  5.0M   0% /run/lock
tmpfs           248M     0  248M   0% /sys/fs/cgroup
eric@cctv:~/bookExample$ █
```

Figure 2.14 df *command with parameter* –h, *it means display for human read. vda1 is the main partition, it still remains 1.9 GB. For biocomputing, 9% available space is not enough*

After finding out the maximum disk space you have, you may need to find out the demands of your program for space. Typically, we would also compute the time needed to complete a job.

First, we may want to make a quick trial run of a smaller dataset for example, and test how long that takes. Then run a larger dataset and then plot a mental picture of how well the size scales in terms of time needed. Take a look also at the sizes of all the temporary files created, some of which may appear in your current working directory, and some perhaps, somewhere else in /tmp or some other swap space, depending on the program's specification. Other useful techniques include using the time command to time how long it takes to run the program.

Command: wc
Action: word count

This command can count the number of lines using the -l flag, or the number of characters -c. wc -l filename will tell you how many lines there are.

Command: head
Action: print the top few lines of a file

For example, head -1000 filename will send 1000 lines of the file to the standard output.

```
Command: tail
Action: print the last few lines of a file
```

For example, `tail` `-20` filename will send the last 20 lines of a file to the standard output.

```
Command: split
Action: split a file into smaller pieces
```

This command can split a file into smaller chunks based on bytes size (for example `-b` `1024` means split into 1 Kb chunks) or based on number of lines (`-l` `20` means split into 20 line chunks).

So this command will split a large file into 100 line chunks and name them with a prefix genefrag; `split` `-l` `20` `genome1001` `genefrag` and this will generate genefragaa genefragab genefragac, etc., each of 20 lines.

```
Command: csplit
Action: split a file into chunks based on a matching
        pattern
```

```
Command: du
Action: disk usage to compute how much disk space is
        used by a files in a folder
```

```
Command: time
Action: when prefixed to a command line command, it
        will carry out the command and time the
        duration of the operation and output the
        time taken as well as the original output
```

More shortcuts if a program is running a long time and you cannot wait, you might want to abort it using `Ctrl-C` as mentioned earlier. But you could suspend it temporarily using `Ctrl-Z` to bring up a command prompt. You could then push the command into the background for running using the `bg` command. After doing something else on the command prompt, you could bring the program which may still be running back to the foreground using the `fg` command. Typically for running a program with lots of output printed to the terminal screen, you could suspend the output to the screen using `Ctrl-S` and resume using `Ctrl-Q`. Alternatively you might like to redirect the output to a file using a powerful concept in UNIX call redirection.

UNIX redirection and pipes

Every UNIX program has a standard input, a standard output and standard error. By default the standard output and standard error channels print to the terminal screen. You could redirect it in the following manner.

Say you have a bigprogram processing a bigdatafile with lots of output. You could redirect the output to another file using the > redirector.

```
bigprogram bigdatafile > bigoutput
```

If you wish to time the program, you could prefix with the time command.

```
time bigprogram bigdatafile > bigoutput
```

This will produce the output of bigprogram working on the bigdatafile and direct it and save it into the file called bigoutput, but at the same time, show how much time it took at the terminal screen.

Sometimes, a bigprogram might process something line by line, and so you need to find out how long it takes.

```
time head -10 bigdatafile | bigprogram > /dev/null
```

This means time the whole program bigprogram but run it with input from bigdatafile, but this time, use only the first 10 lines using the head command directing by means of a pipe all the output into the input of bigprogram, and whatever output from bigprogram to redirect it into /dev/null which is a big black hole, which essentially makes the output disappear. The time taken can then be taken note of.

```
time head -100 bigdatafile | bigprogram > /dev/null
time head -1000 bigdatafile | bigprogram > /dev/null
time head -10000 bigdatafile | bigprogram > /dev/null
```

Each command will give you the time elapsed. From this you can see roughly how long your program will take if you progressively increase the input file size until the total size of the bigdatafile. ls -lh will tell you the size of the bigdatafile, or else, count the number of lines it contains using:

```
wc -l bigdatafile
```

Suppose you ran a program but it looks like it will take five hours. You can abort it by using Ctrl-C and repeat the command to take a smaller number of lines.

UNIX redirection and pipes are one of the most powerful features you can find. The concept of being able to redirect the output of one program to the input stream of another program and chaining them up is a concept worth grasping. More will be discussed in a section in the Appendix.

> Command: top
> Action: list out the top processes

Those of us accustomed to using Windows must know of the famous three finger command Ctrl-Alt-Del. Pressing this and selecting the Task Manager will show up all the processes currently running. Similarly, UNIX has the same capabilities and more. For example, simply enter the command like top. This command displays the current running program, and how much of the processor (CPU) and memory (MEM) is consumed.

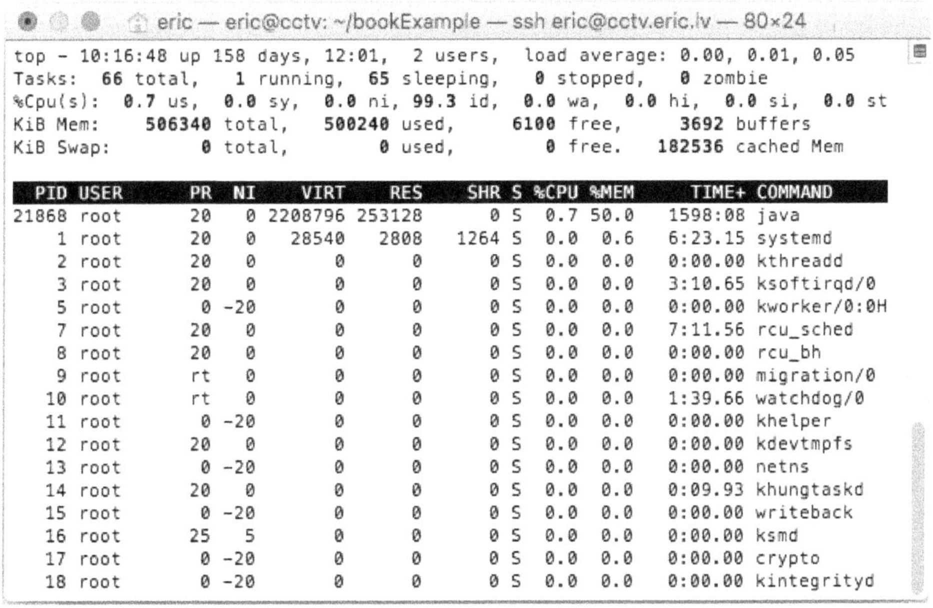

```
● ● ●      eric — eric@cctv: ~/bookExample — ssh eric@cctv.eric.lv — 80×24

top - 10:16:48 up 158 days, 12:01,  2 users,  load average: 0.00, 0.01, 0.05
Tasks:  66 total,  1 running,  65 sleeping,   0 stopped,   0 zombie
%Cpu(s):  0.7 us,  0.0 sy,  0.0 ni, 99.3 id,  0.0 wa,  0.0 hi,  0.0 si,  0.0 st
KiB Mem:   506340 total,   500240 used,    6100 free,    3692 buffers
KiB Swap:       0 total,       0 used,       0 free.  182536 cached Mem

  PID USER      PR  NI    VIRT    RES    SHR S %CPU %MEM     TIME+ COMMAND
21868 root      20   0 2208796 253128      0 S  0.7 50.0  1598:08 java
    1 root      20   0   28540   2808   1264 S  0.0  0.6   6:23.15 systemd
    2 root      20   0       0      0      0 S  0.0  0.0   0:00.00 kthreadd
    3 root      20   0       0      0      0 S  0.0  0.0   3:10.65 ksoftirqd/0
    5 root       0 -20       0      0      0 S  0.0  0.0   0:00.00 kworker/0:0H
    7 root      20   0       0      0      0 S  0.0  0.0   7:11.56 rcu_sched
    8 root      20   0       0      0      0 S  0.0  0.0   0:00.00 rcu_bh
    9 root      rt   0       0      0      0 S  0.0  0.0   0:00.00 migration/0
   10 root      rt   0       0      0      0 S  0.0  0.0   1:39.66 watchdog/0
   11 root       0 -20       0      0      0 S  0.0  0.0   0:00.00 khelper
   12 root      20   0       0      0      0 S  0.0  0.0   0:00.00 kdevtmpfs
   13 root       0 -20       0      0      0 S  0.0  0.0   0:00.00 netns
   14 root      20   0       0      0      0 S  0.0  0.0   0:09.93 khungtaskd
   15 root       0 -20       0      0      0 S  0.0  0.0   0:00.00 writeback
   16 root      25   5       0      0      0 S  0.0  0.0   0:00.00 ksmd
   17 root       0 -20       0      0      0 S  0.0  0.0   0:00.00 crypto
   18 root       0 -20       0      0      0 S  0.0  0.0   0:00.00 kintegrityd
```

Figure 2.15 *Enter the top of the command, the entire screen is constantly refreshing, based on the top processes consuming the machine's resources at any one moment. This will help us understand the system's current working condition. Press q to leave this mode*

Command: ps

Action: print out all processes belonging to the user. ps -ef will print out all processes on the machine. You can identify the process that is causing a problem and note down the process number

Command: kill

Action: kill a process. kill -9 12345 will totally abort a program running with a process number 12345

2.6 Other Useful Commands

Here are a few commands I find quite useful for bioinformatics data operations that I have had the opportunity to work on. These commands plus those given earlier in this chapter, lay the foundation for us to cope with the operations behind the practical aspects of future chapters.

Command: cat

Action: catenate contents to the standard output, which essentially displays the content of a file

```
eric — eric@cctv: ~/bookExample/rawData/raw — ssh eric@cctv.eric.lv — 80×24
[eric@cctv:~/bookExample/rawData/raw$ cat seq1.fastq
@HWI-EAS209_0006_FC706VJ:5:58:5894:21141#ATCACG/1
TTAATTGGTAAATAAATCTCCTAATAGCTTAGATNTTACCTTNNNNNNNNNNNTAGTTTCTTGAGATTTGTTGGGGGAGAC
ATTTTTGTGATTGCCTTGAT
+HWI-EAS209_0006_FC706VJ:5:58:5894:21141#ATCACG/1
efcfffffcfeefffcfffffffddf`feed]`]_Ba_^__[YBBBBBBBBBBBRTT\]][]dddd`ddd^dddadd^BBBB
BBBBBBBBBBBBBBBBBBBBB
eric@cctv:~/bookExample/rawData/raw$
```

Figure 2.16 *To display file contents, the command is:* cat <file name>. *For example,* cat seq1.fastq, *showing all the contents of this sample file*

`cat seq1.fastq` displays of files of this nature by printing the text sequence contents of the file. However, please note the type of content being displayed. When you "`cat <sequence File>`", the run time should be pretty fast for all the text in the file to be output to the terminal screen. However, if the file is a binary file containing an executable program, or a picture, the duration will be long and the binary output may even crash your terminal application. To stop the output, you can try at least two methods. Firstly, press `Control` key then `c`, i.e. `CTRL-C` to abort the execution. You typically have to do this several times. Secondly, you may like to pause the program using `Ctrl-Z` and bring up the command prompt when the program is suspended. At the command prompt, enter `ps -ef | grep cat` to pull up all the programs running `cat`. Identify the correct command, and its corresponding process number, say 12345, then run the command

```
kill -9 12345
```

This will stop and abort the application. The third way is to `CTRL-S` to suspend the output to screen, before executing method 2. Typically, if the screen starts filling up with gibberish because you have just `cat` a non-text file, you have to quickly hit `Ctrl-C` to abort.

```
Command: wget
Action:  connect to a web address and retrieve the
         contents, e.g. retrieve a web page and save
         as an HTML file. To download files from a
         network source, the command to use is "wget
         <complete URL>." This behaves like a text-
         only interface non-graphical Web browser!
```

```
● ◎ ●      ⌂ eric — eric@cctv: ~/bookExample — ssh eric@cctv.eric.lv — 80×24
[eric@cctv:~/bookExample$ wget ftp://ftp.ccb.jhu.edu/pub/data/bowtie_indexes/hg19
.ebwt.zip
converted 'ftp://ftp.ccb.jhu.edu/pub/data/bowtie_indexes/hg19.ebwt.zip' (ANSI_X3
.4-1968) -> 'ftp://ftp.ccb.jhu.edu/pub/data/bowtie_indexes/hg19.ebwt.zip' (UTF-8
)
--2015-11-21 10:20:35--  ftp://ftp.ccb.jhu.edu/pub/data/bowtie_indexes/hg19.ebwt
.zip
           => 'hg19.ebwt.zip'
Resolving ftp.ccb.jhu.edu (ftp.ccb.jhu.edu)... 128.220.233.225
Connecting to ftp.ccb.jhu.edu (ftp.ccb.jhu.edu)|128.220.233.225|:21... connected
.
Logging in as anonymous ... Logged in!
==> SYST ... done.     ==> PWD ... done.
==> TYPE I ... done.   ==> CWD (1) /pub/data/bowtie_indexes ... done.
==> SIZE hg19.ebwt.zip ... 2822736839
==> PASV ... done.     ==> RETR hg19.ebwt.zip ... done.
Length: 2822736839 (2.6G) (unauthoritative)

hg19.ebwt.zip          0%[                        ]   2.06M   397KB/s   eta 3h 1m
```

Figure 2.17 *Actual usage of the wget command to download from UCSC Human Genome website a complete reference sequence hg19 index file. The command output will display the progress of the current download and its estimated completion time. Supported protocols for wget include http and ftp*

Command: scp
Action: secure file copy to or from a remote server

This command allows the user to transfer files between each server, eliminating the need to first save the file to the terminal computer, and then transfer files from the terminal server to another server in an additional step. Bioinformatics files are often very large, so we should not waste bandwidth! Complete instructions are:

scp localfile filename@remote-server-address:full-path-filename

For example, if we want to copy a file called cuda4.cu in the current folder, to a remote server, cueda.eric.net.tw, logging in as user eric, and storing the exact copy as bookExample, in the full directory path /home/eric/bookExample

scp cuda4.cu eric@cuda.eric.net.tw:/home/eric/bookExample

Enter the password corresponding to the remote account userid and the transfer is started.

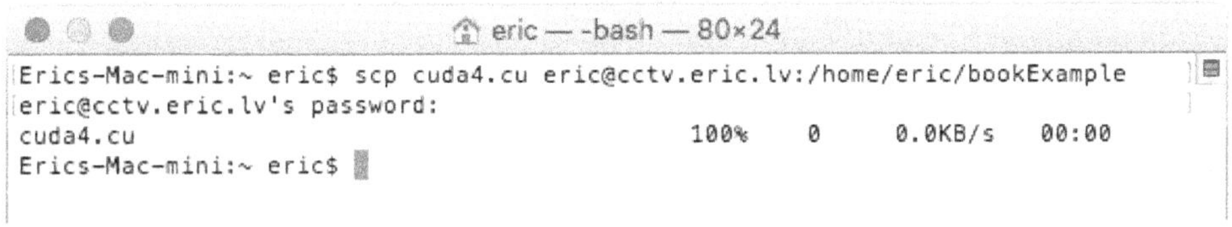

```
● ◎ ●                    ⌂ eric — -bash — 80×24
Erics-Mac-mini:~ eric$ scp cuda4.cu eric@cctv.eric.lv:/home/eric/bookExample
eric@cctv.eric.lv's password:
cuda4.cu                                 100%    0     0.0KB/s    00:00
Erics-Mac-mini:~ eric$ ▓
```

Figure 2.18   *To put this local file cuda4.cu into the remote machine and save into a remote host speci-*
fied full directory path and file name corresponding to the permissions of the login user, and key in the
remote host password to initiate the transfer


```
Command: tar
Action: wrap up a series of files in a folder into a
        single file
```

When you downloaded files from the Internet, especially software packages, many of them come in a compressed and encapsulated form, typically with a "tar.gz". This means that all the files are compressed with GNU zip after the folders and all the files are packed up using the **tar** command. This was done using the command:

```
tar -zcvf file.tar folder
```

It will **tar** up all files in the folder into a file called file.tar and then zip it up into file.tar.gz.

To decompress and untar, use the command:

```
tar -zxvf file.tar.gz
```

The **tar** command originally refers to tape archive.

Typically, a packaged folder will generate a file with a file name suffixed with the .tar extension. If it is subsequently compressed in gzip format, another extension .gz will appear at the end. So if it is bzip2 compressed format, .bz2 will be appended (so if you need to compress c, in z or unbzip2, j, do add the parameters such as -czvf or -cjvf respectively). If you wish to unzip/untar such files, you would use -zxvf or -jxvf as the **tar** parameter.

For example, to package all files in the folder, rawData, and then compress it in gzip format, we will get as output a file called rawData.tar.gz. So the complete command is:

```
tar -zcvf rawData.tar.gz rawData/
```

To unzip and untar, the command is:

```
tar -zxvf rawData.tar.gz
```

And it will unpack everything in the rawData/ folder.

```
● ● ●      🔒 eric — eric@cctv: ~/bookExample — ssh eric@cctv.eric.lv — 80×24
[eric@cctv:~/bookExample$ tar -czvf rawData.tar.gz rawData/
rawData/
rawData/raw/
rawData/raw/seq1.fastq
rawData/raw/seq1.fastq.bak
[eric@cctv:~/bookExample$ ls
cuda4.cu  hg19.ebwt.zip  rawData  rawData.tar.gz  runWorkflow.sh
[eric@cctv:~/bookExample$ tar -cjvf rawData.tar.bz2 rawData
rawData/
rawData/raw/
rawData/raw/seq1.fastq
rawData/raw/seq1.fastq.bak
[eric@cctv:~/bookExample$ ls
cuda4.cu          rawData              rawData.tar.gz
hg19.ebwt.zip  rawData.tar.bz2  runWorkflow.sh
eric@cctv:~/bookExample$ ▌
```

Figure 2.19 *For the rawData archive, the actual commands for packing is* tar *and compressed with gzip or bzip2 using the appropriate flags* -z *or* -j. *As for decompression, it is also very easy, just use the* -x *flag instead of* -c *(for compress):* tar -xvf <packaged file name>

Then the **tar** command will put the files into a folder called by the name of the file without the extensions. In the example screenshot, we demonstrate that the directive applies equally well to decompression of gz or bz2 files. Finally, if you get a zip format file to decompress input unzip the ZIP file name.

```
● ● ●      🔒 eric — eric@cctv: ~/bookExample — ssh eric@cctv.eric.lv — 80×24
[eric@cctv:~/bookExample$ ls
cuda4.cu          rawData              rawData.tar.gz
hg19.ebwt.zip  rawData.tar.bz2  runWorkflow.sh
[eric@cctv:~/bookExample$ tar -xvf rawData.tar.gz
rawData/
rawData/raw/
rawData/raw/seq1.fastq
rawData/raw/seq1.fastq.bak
[eric@cctv:~/bookExample$ tar -xvf rawData.tar.bz2
rawData/
rawData/raw/
rawData/raw/seq1.fastq
rawData/raw/seq1.fastq.bak
[eric@cctv:~/bookExample$ ls
cuda4.cu          rawData              rawData.tar.gz
hg19.ebwt.zip  rawData.tar.bz2  runWorkflow.sh
eric@cctv:~/bookExample$ ▌
```

Figure 2.20 *In this example, we show how unzip and untar process in the stated folder, and how unzip/untar of rawData.tar.gz or rawData.tar.bz2 generates the resulting folders, which* ls *lists out for us to visually inspect what has taken place. Note that the unzipping and untarring process will actually clobber any pre-existing folders and files, i.e. overwrite all pre-existing files*

CHAPTER 3

Checking Sequence Quality

3.1 Basic High-throughput Sequencing

DNA sequencing is the term given to the process of determining the order of the four nucleotide bases A, T, C and G, which compose the genes of the genome of an organism. As early as the 1950s, Englishman Fred Sanger (one of only four individuals in the history of the Nobel Prize to have won the Prize twice; once in 1958 for sequencing proteins and in 1980 for the dideoxy method of DNA sequencing) began to study how to determine the sequence of biological polymers. By the 70s, he successfully invented the technique of biochemically determining the sequence of DNA, which eventually led to the start of the Human Genome sequencing project in the 90s. Sanger's method of sequencing DNA (the dideoxy chain termination method) was accurate, but cost was high in terms of large amounts of chemicals needed for the biochemical reactions, and combined with the polymerase chain reaction (PCR) to complete the sequencing process, it was very time consuming.

By the 1990s, the world was anxiously awaiting the completion of the sequencing of the entire human genome. As far as the scientists were concerned, as soon as the human genome is completed, they believed that we can obtain a much better grasp of the molecular basis of disease and its potential cure. The US government and the public sector began to compete in the completion of the human genome sequencing.

In the process of this friendly competition, high performance sequencing techniques were developed very rapidly. High throughput sequencing also known as Next Generation Sequencing (NGS), adopted a slightly different approach to traditional methods of sequencing. The general idea was to take the entire genome and fragment them into short segments for biochemical sequence determination, and using laser-assisted reading of the different nucleotide bases, A, T, C or G in an automated manner using computers.

Subsequently, these trillions of short sequences had to be assembled by computer techniques in order to complete the entire three billion base pairs of the human genome sequence. This technique also known as shot-gun sequencing, although it is very simple in approach, in reality, the process of sequence assembly was a major challenge, and some of the then unsolvable problems were left to the scientific community to resolve. In the same manner, this technique also known as high throughput sequencing, has become a global race in the genome sequencing industry, to see who can come up with the fastest and least expensive process. For example, next generation sequencing companies including Illumina (merged with Solexa), Roche (acquired 454 Life Science), ABI and Helicos have independently developed ever increasingly superior sequencing technologies, with their platforms achieved even faster rates of sequencing, longer sequence read-lengths, and most importantly higher accuracy. In this way, they were able to gradually achieve much lower costs of sequencing without compromising sequencing accuracy, aiming for the ideal of bringing the cost of sequencing an entire human genome down to US$1,000 so that anybody can have the chance of having their own genome fully sequenced, thereby accelerating their chance of getting prescribed personalized medicine.

The genomics industry's intense efforts to develop better sequencing techniques, is largely driven by the goal of achieving better cure for human disease. One idea was to gather as many human whole genome sequences as we can. As a baseline, most of these whole genomes would have come from normal healthy individuals, but as soon as a person with a specific disease is clearly identified, we can compare that person's genome with those of healthy ones, to determine where the differences in the genome sequence are. Based on these differences, we can further investigate the causative mechanism at the genetic level, so as to better understand how the disease comes about, as well as to identify the potential sites of drug intervention. By gathering such large volumes of genomic sequence data, we may possibly attain the goal of predicting disease based on identifying mutations in the genome. In the past, there have been a large body of data correlating chromosomal mutations with human genetic disease. If we can pinpoint the exact location of these mutations down to the individual nucleotide bases, we can establish a stronger scientific basis for predicting disease. In fact, for those of us who are interested, there is already a large body of data and many databases accessible via the Internet, which contain the latest scientific and medical findings to date.

Over the past few years, high throughput genome sequencing technology and biomedical research in disease, has resulted in a new development, that of Exome Sequencing. Instead of sequencing the entire genome, by focusing on only sequencing the specific regions of the genome responsible for producing proteins, we may be able to reduce the costs and cut down the time for diagnosis. Exomes are the portions of the genome which code for proteins. Approximately 30 million bases (about 1%) of the over three billion bases of the whole human genome code for exons. These exons are the bits of sequences in the genome which code for proteins. Mutations in the genome sequence results in mutations in the proteins, resulting in defective function, which may lead to disease. By concentrating on only 1% of the genome, we can cut cost and reduce time to the diagnosis of any human disease that is caused by mutation.

High performance sequencing is still an unending race. The latest next generation sequencing technique has already been commercialized. For example, the technique of the company Nanopore, basically can achieve the sequencing of a single molecule. By passing a DNA molecule into a tiny pore, and measuring the electrical changes as the DNA passes through the pore, they are able to distinguish between difference bases, thereby elucidating the genome sequence in an even more high throughput manner and reducing the need for the use of chemical reactions, and obviating the need for laser excitation of the chemicals. In this way, the chemical luminescent technologies are progressively giving way to a new generation of technologies that rely directly on electrostatic determination of the individual bases of the DNA sequence.

3.2 Challenges of High Throughput Genome Sequencing

In the earlier section, we have alluded to a number of problems intrinsic to high-throughput sequencing. The biochemical reactions involved and the reagents used are intrinsically error-prone, and consequently, errors in the sequencing results are inevitable. Every reagent deployed in the current sequence elucidation process has been unable to avoid such errors.

Current sequencing platforms rely on chemical synthesis, followed by irradiation of the fluorophores with laser light. A camera immediately captures the fluorescence generated and records pictures for image recognition. Image recognition algorithms have no way of recognizing 100% any grey areas between any two kinds of bases. Even if the errors arising from the chemical experiments can be avoided, the technical computational problems have to be solved too.

Shot-gun sequencing generates sequences of short fragments. Given that each fragment may be 75–400 nucleotides long (depending on the various sequencing platforms), when compared to three billion bases of a human genome, there may be hundreds of thousands of sequences with all kinds of overlaps. How can a computer program take all these overlapping sequences and assemble these fragments into the complete genome? And if the sequencing machine itself generates errors in the overlaps, even the best computer programs may not be able to overcome erroneous assembly easily.

With so many short sequences, the algorithm used to assemble requires efficient use of memory. With too much data to process, with each sequence having an element of error, even the completion of the preliminary results of the assembly is not easy. Many computer scientists dedicated to developing algorithms to handle high throughput sequencing data today have come up with solutions of differing technical computational requirements and differences in speed and accuracy.

So the whole exercise of sequencing a genome has become not so much a deterministic exercise as a probabilistic guess. The credibility of such a guess is very much in question, often reduced to a probabilistic score. For reasons of interoperability and cross comparability, a standardized score has become important, to the extent that any published data can be evaluated against that standard score, and any scientist can ascribe a level of trust to the accuracy and quality of the data.

3.3 Standards of Quality Score

As early as 1992, a concept of quality was already established for sequencing. By 1995, Bonfield and Roger Staden published a logarithmic error rate based on the concept of evaluation criteria called the Phred Score. This has been adopted by the various sequencing platforms as a standard of interoperability of data, and as a method of comparing the accuracy between different datasets. To encapsulate the quality score, a FASTQ format has also been adopted as a standard format.

If you get a bunch of sequences in FASTQ format, in addition to the sequence data, there is a Quality Score. A basic FASTQ file can be opened with a text editor or displayed using the `cat` command:

```
@SEQ_ID
GATTTGGGGTTCAAAGCAGTATCGATCAAATAGTAAATCCATTTGTTCAACTCACAGTTT
+
!''*((((***+))%%%++)(%%%%).1***-+*''))**55CCF>>>>>>CCCCCCC65
```

This FASTQ format may look like there are no scores in them, just like the file format called FASTA, which does not contain any numbers in the DNA sequence data. However, we do note that corresponding to each base in the sequence, in a parallel row, is a character of some kind. While not directly showing scores, each character in the second row corresponding to each base of the DNA sequence actually is the Quality Score represented by characters of the ASCII character encoding, each of which has a corresponding decimal number associated with it.

The so-called ASCII code, refers to the standards of the United States governing the exchange of information (American Standard Code for Information Interchange). It was developed and standardized in the late 1960s as a computer code that encodes common English letters, punctuation marks and special characters such as @, #, etc., giving each a corresponding number in binary, decimal or hexadecimal form. Each time we use the keyboard and strike a character, there is a unique code which gets transmitted to the computer.

Table 3.1 *FASTQ quality score characters FASTQ decimal fractions corresponding to characters in the ASCII character conversion table*

Score	Char	Score	Char	Score	Char	Score	Char	Score	Char	Score	Char
32		48	0	64	@	80	P	96	`	113	q
33	!	49	1	65	A	81	Q	97	a	114	r
34	"	50	2	66	B	82	R	98	b	115	s
35	#	51	3	67	C	83	S	99	c	116	t
36	$	52	4	68	D	84	T	100	d	117	u
37	%	53	5	69	E	85	U	101	e	118	v
38	&	54	6	70	F	86	V	102	f	119	w
39	'	55	7	71	G	87	W	103	g	120	x
40	(	56	8	72	H	88	X	104	h	121	y
41	)	57	9	73	I	89	Y	105	i	122	z

The ASCII code table is typically found in the appendix of any introduction to Computer Science textbook. If you are really curious to check up characters appearing in the above example, ! is 33, * is 42, + is 43, so we should be able to get a feel of how smart the guys, who first thought of using the ASCII encoding scheme to define quality score that matches to a single byte of a sequence base, were. Replaced with a letter, you can save at least half of the amount of data transferred. For a Quality Score with ten-digit decimal display, then we would have to use two numeric characters.

Now that we understand the FASTQ file format and the encoding system of their quality score, next we need to understand the basis for these scores, which simply speaking, is just a formula

$$Q = -10 \log_{10} P$$

or

$$P = 10^{-Q/10}$$

The Q value represents the quality score (Phred Score), P represents the probability of sequencing errors. The lower the P value, the higher the Q value will be. This allows us to represent the quality of sequencing results with a scale of accuracy and reliability. If the Q value is 10, one base of every ten will be incorrectly assigned; the accuracy rate would be 90%. Similarly, a Q value of 30 means that the sequencing process would make a mistake every one thousand bases, giving an accuracy rate of 99.9%. Based on this explanation, we should be able to gauge the scale on the quality score bar. The following table lists the quality score and error rates, and the accuracy of the reference controls.

Table 3.2 *Phred Score Quality Score, relative error rates and accuracy of the relationship*

Quality score (Phred score)	Sequencing error frequency	Accuracy
10	1 error in 10 bases	90%
20	1 error in 100 bases	99%
30	1 error in 1000 bases	99.9%
40	1 error in 10000 bases	99.99%
50	1 error in 100000 bases	99.999%

3.4 Quality Check

Typically, samples sent to biotech high-throughput sequencing laboratories will be analyzed on their proprietary sequencing platform, from which we will obtain the results in the form of FASTQ files. We can process these FASTQ files using software which outputs a summary report on the sequence and displays charts to help determine the quality of the sequencing data. Let us look at two quality analysis software, FastQC and FASTX-Toolkit.

FastQC

The first is FastQC, which is a software developed using Java. It accepts FASTQ, BAM and SAM file formats (the latter two will be described in detail in the subsequent chapter on sequence alignment), and then generates a complete report containing both sequence, statistics and graphical display.

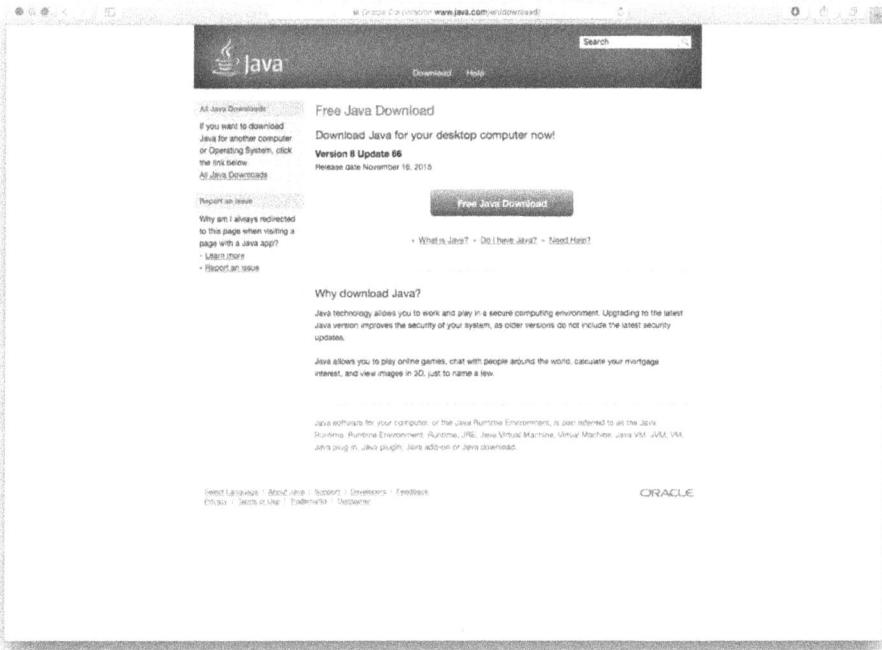

Figure 3.1 *Windows or Mac users who need to install a Java Virtual Machine, can download the JVM from the website, http://java.com, and install the JVM on your Windows computer free of charge*

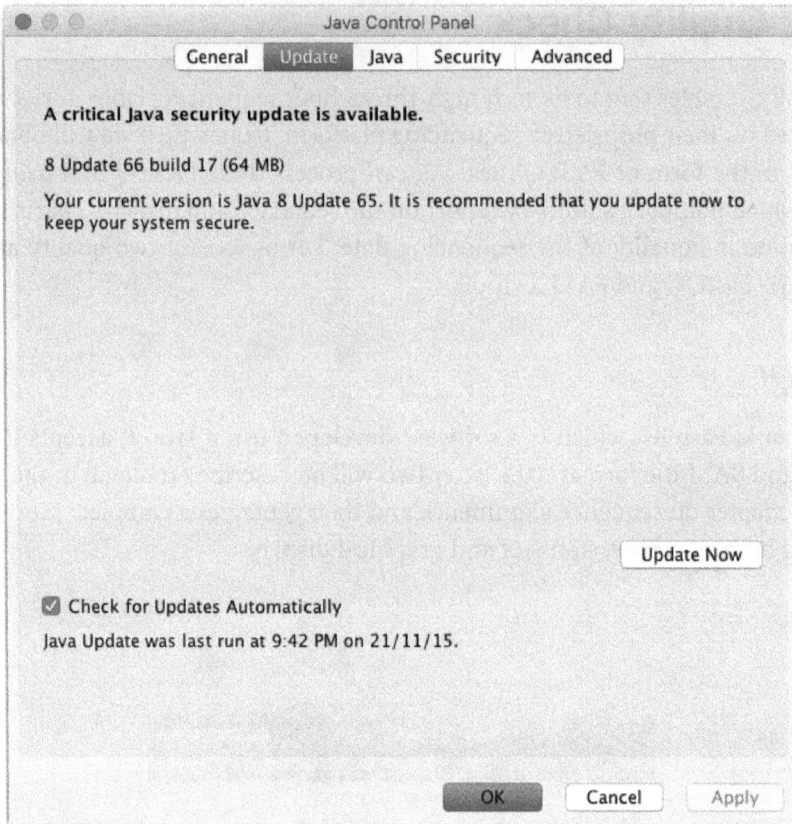

Figure 3.2   *For macOS, you can navigate to the System Preferences to find your Java Preference settings and determine the version of Java*

To process such FASTQ files on your PC or Mac desktop environments, please make sure that you have pre-installed Java Virtual Machine (JVM) on your computer. For the Mac, it is usually already built-in; for Windows users, please visit Oracle's website (http://java.com/) to download and install the JVM.

Then navigate to FastQC official website to download the appropriate version of FastQC. (http://www.bioinformatics.babraham.ac.uk/projects/download.html#fastqc). Please ensure that the version number is at least FastQC v0.11.4 (current at the time of writing) and download the version appropriate to your computer's operating system, namely Windows/Linux operating system tarball (Win/Linux zip file) or Mac-specific disk image file (Mac DMG image).

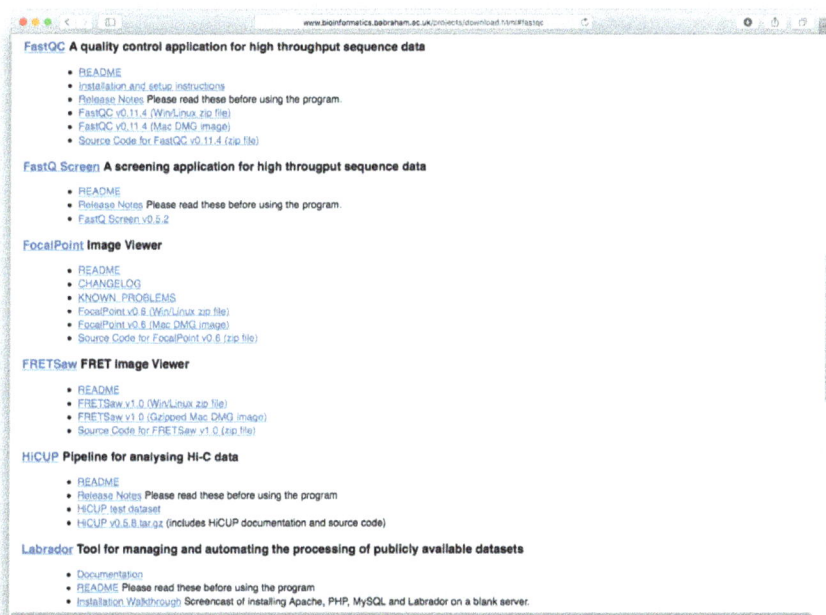

Figure 3.3   *Please be careful to make sure you download the correct version of FastQC appropriate to your own platform*

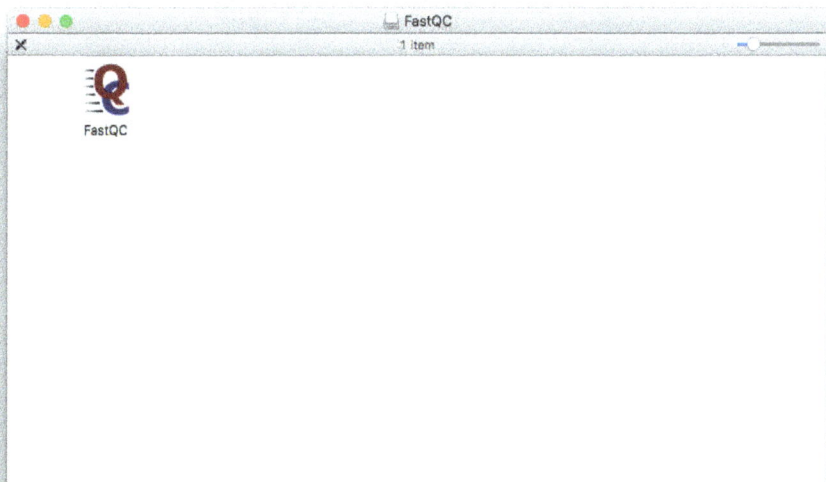

Figure 3.4   *Upon mounting the DMG file using the macOS, the only object inside is FastQC. Double-click on it to execute. The screen interface for installation is very simple. It prompts you to open the sequence file from the menu. Click "File" menu and select "Open"*

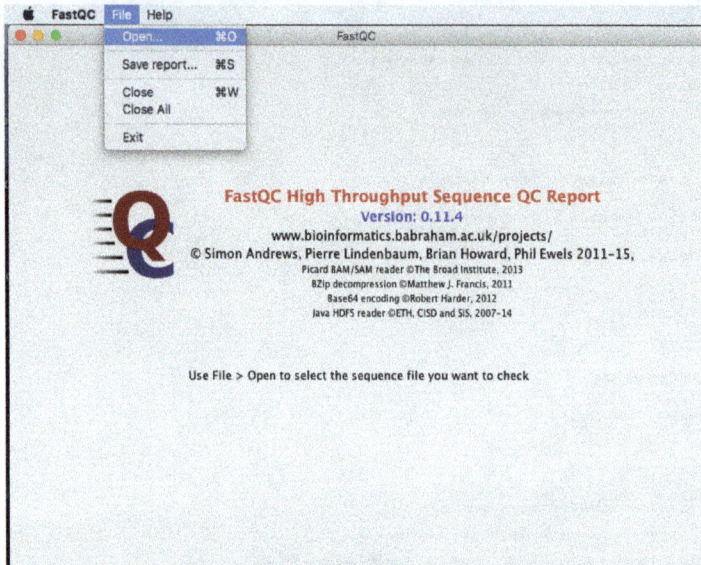

Figure 3.5 *Click FastQC main screen "File" menu, and select the "Open" to open the dialog screen, and choose your own sequence data file. FastQC supports Fastq, Casava Fastq, BAM or SAM file formats. If the Fastq file extension is* txt, *remember to choose "All Files" so that you can see the file. If there are too many files in the folder, you should specify a specific file format as a suitable filter to remove irrelevant files from the selection box*

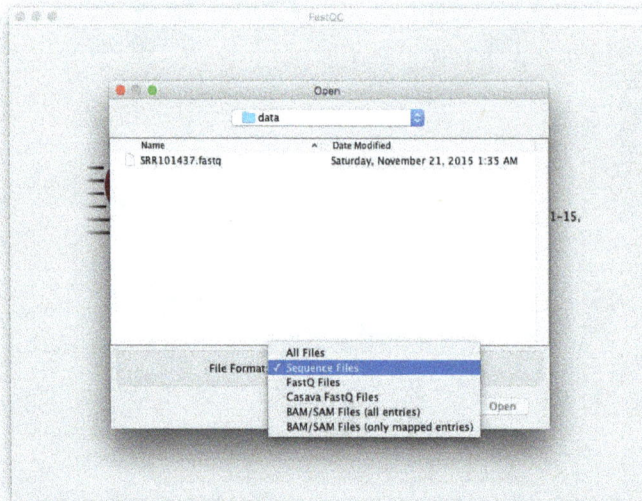

Figure 3.6 *Open the sequence file window, and select the sequence quality files to be analyzed. After loading the sequence, FastQC will read the file and automatically analyze the file. Depending on the sequence file size, and the performance of your computer hardware, it may take a longer time to complete the analysis, so please be patient. You can see the current progress of the program while it is processing the data in central pane of the FastQC application*

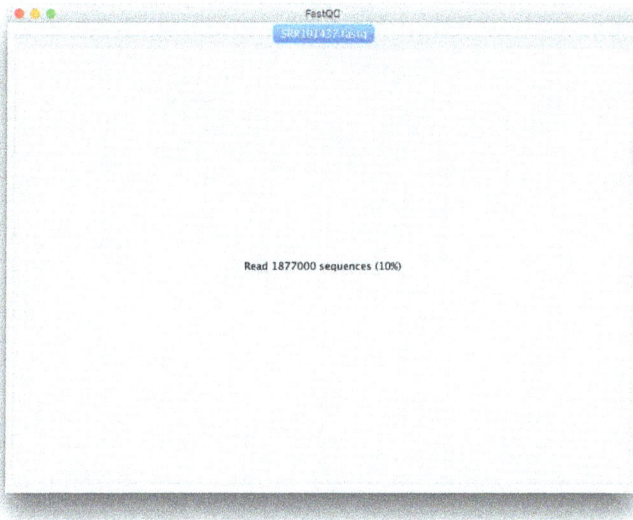

Figure 3.7 *When FastQC starts analyzing a sequence file, from the main screen display, the progress can be monitored. Once the analysis is completed, the screen will show a summary of the results, including the sequencing platform, total number of sequences, sequence length and GC content ratio*

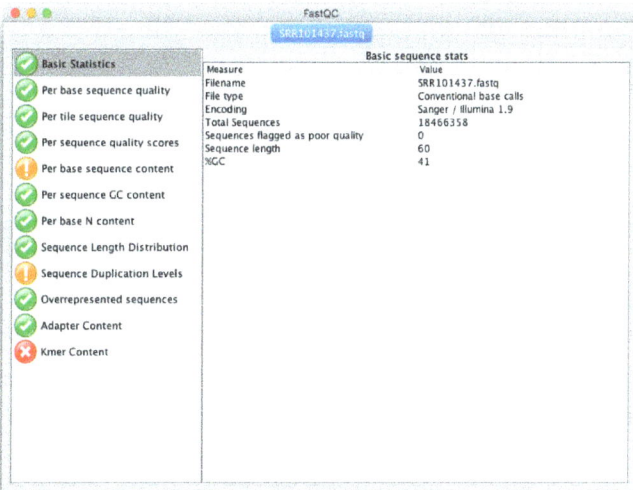

Figure 3.8 *After the analysis is completed, the left pane of the FastQC window will contain the details of each analysis. Although it is self-explanatory, and very simple to use, but note that the results can also be packaged into a single zip file, whether you need the result in a chart, or not, you can "*Save Report*" from the dropdown File menu. To save the report is very simple, create a suitable file name and click on the save button, and then a zip file will be created*

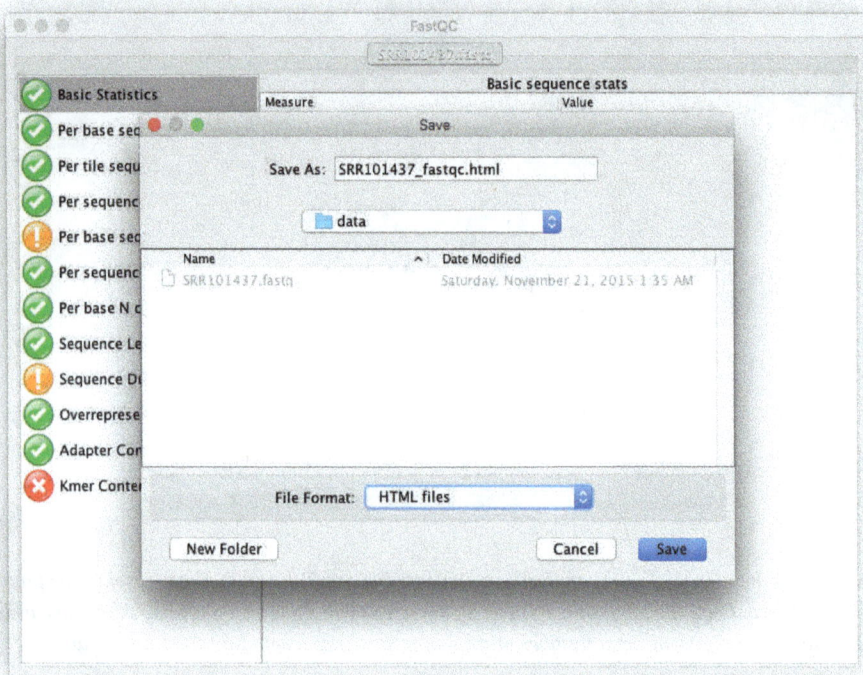

Figure 3.9 *When you wish to save the report, enter a suitable file name in order to save the entire sequence analysis report as a HTML file*

When you expand the compressed file of the sequence analysis report, you can see that there are two txt files, fastqc_data.txt and summary.txt. The main information is in the first file, where you may need to use Excel to visualize or re-draw the data and to allow you to determine the quality. In the Images folder of the entire sequence analysis package, the results come in the form of a variety of charts.

To view the report, please open fastqc_report.html file using a Web browser to navigate all the different sections and graphs of the report. As a reminder, if you want to submit this report, please remember not to miss out any file in the package.

While FastQC may be a user-friendly graphical interface, it can only handle limited file sizes. When your sequence data is very large, especially those stored in a Linux server environment, you will need to use FASTX-Toolkit in order to manage such large datasets.

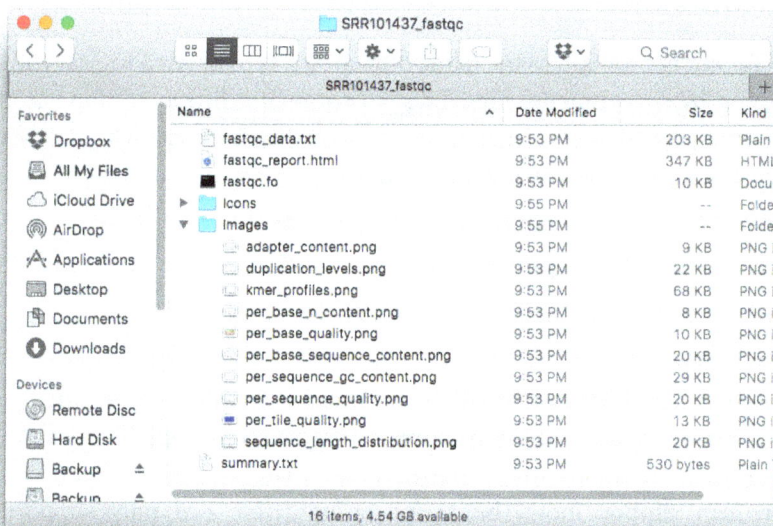

Figure 3.10 *After uncompressing the FastQC packaged compressed file, you can quickly glance through the entire results via the html file*

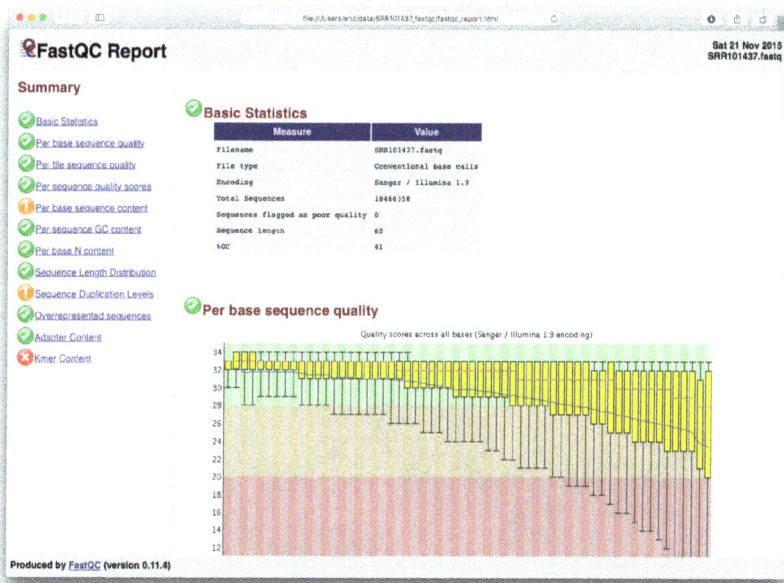

Figure 3.11 *Viewing the HTML file gives a very similar interface to the FastQC output screen*

FASTX-Toolkit

FASTX-Toolkit is a toolkit, as the name indicates, it is to be used to deal with a class of FASTA or FASTQ sequence files. This is a command-line operation based applications package. It is used to analyze the data quality fastx_quality_stats statistical tools. This toolkit can be downloaded free of charge from its official website, and I will guide you on how to install the correct version of the software.

Which version of the computer architecture should you choose? This is something we need to discuss. For those unfamiliar with computer systems, the latest 64-bit architecture processors, such as Intel Core 2 Duo, Core i5, Core i7, Xeon and AMD Athlon 64, are currently the mainstream processors for computers which, by the way, are also backwards compatible with the 32-bit older architecture. However, the opposite is not true, a 32-bit machine cannot run 64-bit applications. As for the operating system of the computer, there is also a 64-bit or 32-bit version, so a 64-bit hardware can support both, but a 32-bit hardware can only support 32-bit

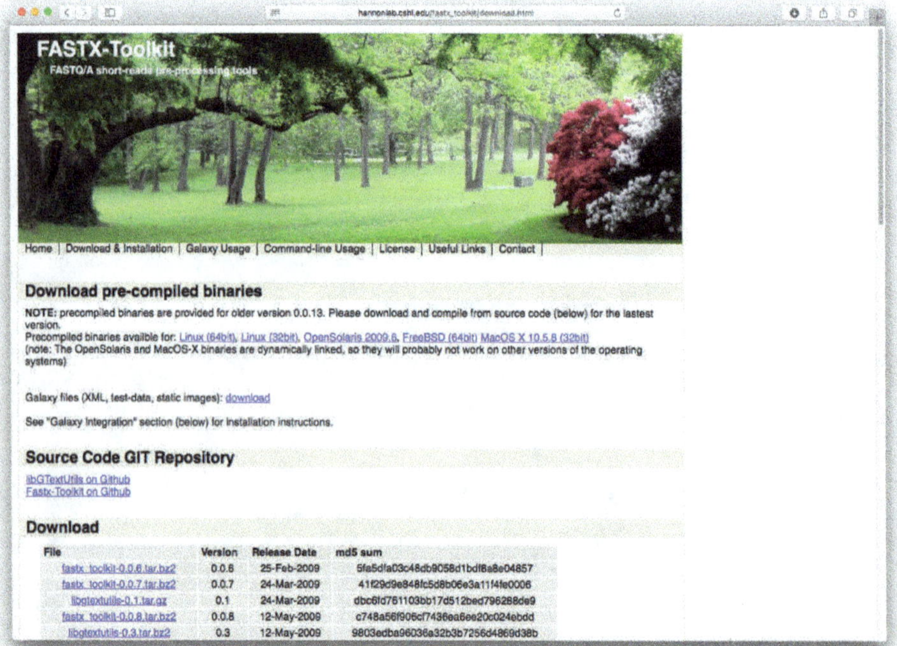

Figure 3.12 *FASTX-Toolkit download site: http://hannonlab.cshl.edu/fastx_toolkit/download.html. In the FASTX-Toolkit download webpage, from the Download options of pre-compiled binaries, and several versions of the operating system, please select Linux, 64-bit or 32-bit version, depending on your PC*

operating system. Generally speaking, a 64-bit computer would have come pre-installed with a 64-bit version operating system.

In the design of software, the computer needs to be programmed to use the system correctly and efficiently during execution of the software. Each software application is generated by the compilation of its source code to generate the executable binaries, which are machine readable codes. A computer programmer would be able to read the contents of the source code and thus understand how it is written and what needs to be adjusted to operate on a 32-bit or 64-bit architecture. Generally, in order for the program to achieve its best performance, the programmer will source for compilers for a variety of system platform. Each compiler will be designed for the corresponding processor architecture with features to facilitate optimization of the code and to enhance the applications run-time performance.

This principle sounds simple, but in practice it is a great challenge. First, we have to establish a comprehensive compilation environment, with all the libraries called from the source code, to be installed and ready for compilation. During the process of compilation, if a desired library is missing, the program will fail to compile.

So, the process of compilation from the source code, is not really a trivial task. In view of this situation that not everyone with biological competence has the full ability to handle this process, developers of software will typically perform the compilation for several common operating system platforms. Once successfully completed, they will provide the compiled binary machine-readable code, i.e. the executable program called "pre-compiled binaries" for end-users to simply download the correct version of the program binary code that is compatible with their operating system and hardware architecture for the best performance.

Our demonstration platform is a 64-bit Ubuntu, and so I downloaded Linux 64-bit compatible software. Remember the `wget` command? Copy the selected software, download URL web address directly into the command line terminal to retrieve the software. After downloading the files, ensure that they are completely downloaded and not paused half way. Extract the tar-compressed file with the `tar -zxvf` parameters. Once completed, you will see a folder named bin, fastx_quality_stats.

In addition, a well-designed software will typically be foolproof, so when the user enters the wrong command or if the syntax is incorrect, it will notify the error and display the help screen automatically.

From the fastx_quality_stats instructional manual, the usage is as follows:

```
fastx_quality_stats -i <fastq file> -o <output file>
```

```
● ◐ ●   ⌂ eric — root@galaxy: ~/bookexample — ssh root@10.12.2.157 — 80×24
[root@galaxy:~/bookexample# tar xfv fastx_toolkit_0.0.13_binaries_Linux_2.6_amd64 ▤
.tar.bz2
./bin/fasta_clipping_histogram.pl
./bin/fasta_formatter
./bin/fasta_nucleotide_changer
./bin/fastq_masker
./bin/fastq_quality_boxplot_graph.sh
./bin/fastq_quality_converter
./bin/fastq_quality_filter
./bin/fastq_quality_trimmer
./bin/fastq_to_fasta
./bin/fastx_artifacts_filter
./bin/fastx_barcode_splitter.pl
./bin/fastx_clipper
./bin/fastx_collapser
./bin/fastx_nucleotide_distribution_graph.sh
./bin/fastx_nucleotide_distribution_line_graph.sh
./bin/fastx_quality_stats
./bin/fastx_renamer
./bin/fastx_reverse_complement
./bin/fastx_trimmer
./bin/fastx_uncollapser
root@galaxy:~/bookexample# ▮
```

Figure 3.13   *After downloading the FASTX-Toolkit, uncompress and extract with the* `tar` `-zxvf` *parameters to the file contents. Explore the installed folders. Try executing* `./fastx_quality_stats -h,` *which means that with the -h (help) flag, the execution of the application fastx_quality_stats will produce a help list of parameters which you can use. Typically a software application under the text user interface will have -h or --help parameter as a standard option to guide and assist new users and old, in the proper use of the program. It displays the software parameters and usually gives a brief explanation of how to perform the computation, what command syntax and input file formats are accepted, etc., and in particular, instructions on how to enter the correct command.*

For example,

> `fastx_quality_stats -i SRR101437.fastq -o output.txt`

In a short while, you will be brought back to the command prompt, and when you browse the directory using `ls`, you will see the appearance of a new file, called output.txt. Display the contents using the `cat` command, and you will see the results of the analysis of this fastq file, including statistics and quality scores at every sequence position of each and every line of the sequence file.

Enter this command

> `./fastq_quality_boxplot_graph.sh -h`

```
● ● ●   eric — root@galaxy: ~/bookexample/bin — ssh root@10.12.2.157 — 80×24
root@galaxy:~/bookexample/bin# ./fastx_quality_stats -h
usage: fastx_quality_stats [-h] [-N] [-i INFILE] [-o OUTFILE]
Part of FASTX Toolkit 0.0.13 by A. Gordon (gordon@cshl.edu)

   [-h]        = This helpful help screen.
   [-i INFILE] = FASTQ input file. default is STDIN.
   [-o OUTFILE] = TEXT output file. default is STDOUT.
   [-N]        = New output format (with more information per nucleotide/cycle)
.

The *OLD* output TEXT file will have the following fields (one row per column):
        column  = column number (1 to 36 for a 36-cycles read solexa file)
        count   = number of bases found in this column.
        min     = Lowest quality score value found in this column.
        max     = Highest quality score value found in this column.
        sum     = Sum of quality score values for this column.
        mean    = Mean quality score value for this column.
        Q1      = 1st quartile quality score.
        med     = Median quality score.
        Q3      = 3rd quartile quality score.
        IQR     = Inter-Quartile range (Q3-Q1).
        lW      = 'Left-Whisker' value (for boxplotting).
        rW      = 'Right-Whisker' value (for boxplotting).
        A_Count = Count of 'A' nucleotides found in this column.
```

Figure 3.14   *Try using the* -h *help parameter to explore the help messages of the program,*
fastx_quality_stats

```
● ● ●   eric — root@galaxy: ~/bookexample/bin — ssh root@10.12.2.157 — 80×24
root@galaxy:~/bookexample/bin# cat output.txt
column  count   min     max     sum     mean    Q1      med     Q3      IQR     l
W       rW      A_Count C_Count G_Count T_Count N_Count Max_count
1       18466358        0       33      589457053       31.92   32      33      3
3       1       31      33      6464834 4489780 3669015 3830048 12681   18466358
2       18466358        0       34      594256344       32.18   32      33      3
4       2       29      34      5427419 2469651 4897043 5672069 176     18466358
3       18466358        0       34      585408769       31.70   32      33      3
4       2       29      34      6148131 3977400 3410266 4930105 456     18466358
4       18466358        0       34      587596508       31.82   32      33      3
4       2       29      34      5751575 3837815 3348686 5527558 724     18466358
5       18466358        0       34      587886713       31.84   32      33      3
3       1       31      34      5518835 3603086 3973529 5370066 842     18466358
6       18466358        0       34      587209887       31.80   32      33      3
3       1       31      34      5439183 3782129 3725756 5519262 28      18466358
7       18466358        0       34      587287527       31.80   32      33      3
3       1       31      34      5614801 3726274 3597453 5527808 22      18466358
8       18466358        0       34      587150177       31.80   32      33      3
3       1       31      34      5105469 3862895 3061985 6435982 27      18466358
9       18466358        0       34      587223516       31.80   32      33      3
3       1       31      34      4827376 3857509 3386179 6395254 40      18466358
10      18466358        0       34      584097054       31.63   31      33      3
3       2       28      34      5075310 3924141 3646472 5820391 44      18466358
11      18466358        0       34      581558802       31.49   31      33      3
```

Figure 3.15   *The output text file from a fastx_quality_stats analysis shows the generated statistics at*
each sequence position. Often it is not enough to study the numbers in the text file output. Like FastQC,
fastx_quality_stats has a fastq_quality_boxplot_graph.sh file. From this program's file name, we can
infer that it will draw a box plot graphical visualization of the output text file

It will display the accepted parameters and their effects on the program, for example, the -i parameter represents the input file name, which refers to the fastx_quality_stats statistical results, while -o parameter will allow you to state the output image file name you wish to save into.

If you want to add titles to the boxplot chart, please use the -t parameter plus the title. Default output image format is PNG, unless you wish to output in PostScript format, please use the -p parameter. Let us take a look at the following example:

```
./fastq_quality_boxplot_graph.sh  -i  output.txt  -o
output.png -t SRR101437
```

```
● ◉ ● ⌂ eric — root@galaxy: ~/bookexample/bin — ssh root@10.12.2.157 — 80×24
[root@galaxy:~/bookexample/bin# ./fastq_quality_boxplot_graph.sh —h
Solexa-Quality BoxPlot plotter
Generates a solexa quality score box-plot graph

Usage: ./fastq_quality_boxplot_graph.sh [-i INPUT.TXT] [-t TITLE] [-p] [-o OUTPU
T]

    [-p]            - Generate PostScript (.PS) file. Default is PNG image.
    [-i INPUT.TXT]  - Input file. Should be the output of "solexa_quality_statistic
s" program.
    [-o OUTPUT]     - Output file name. default is STDOUT.
    [-t TITLE]      - Title (usually the solexa file name) - will be plotted on the
    graph.

[root@galaxy:~/bookexample/bin# ./fastq_quality_boxplot_graph.sh —i output.txt -o
  output.png —t SRR101437
  root@galaxy:~/bookexample/bin# ▊
```

Figure 3.16 *First use the -h parameter for fastq_quality_boxplot_graph.sh to understand the command instruction and from there generate the output image file accordingly, with title if necessary*

Next, you will need to view the boxplot graphically. You cannot view a picture from the command line. You will need to file transfer the png or Postscript file back to your PC or Mac, using a free file transfer software, like FileZilla, using the same port 22 to connect to this host, and pull back the graphics files. Once retrieved to your home machine, you can use a variety of graphics display software to visualize the box plot.

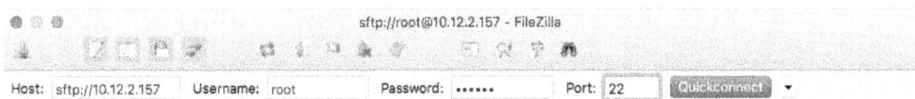

Figure 3.17 *At the top of the main program FileZilla, enter the host address, username, password, and remember to fill in the port number as port 22. Then press the button "Quick Connect" to access the folders in the host*

After connecting with the host, the FileZilla main screen will appear with two split window panes. On the left pane is the view of the current directory folder you are using on your computer, while the right pane displays the folder on your remote Linux host server, which you are logged into. Switch the working directory to the working folder, which contains the output PNG files, download them to your computer from the remote Linux server.

Figure 3.18 *From the right window pane (which represents your remote Linux server), drag and drop the appropriate PNG files into the appropriate panel on your left pane, which is the folder in your local PC*

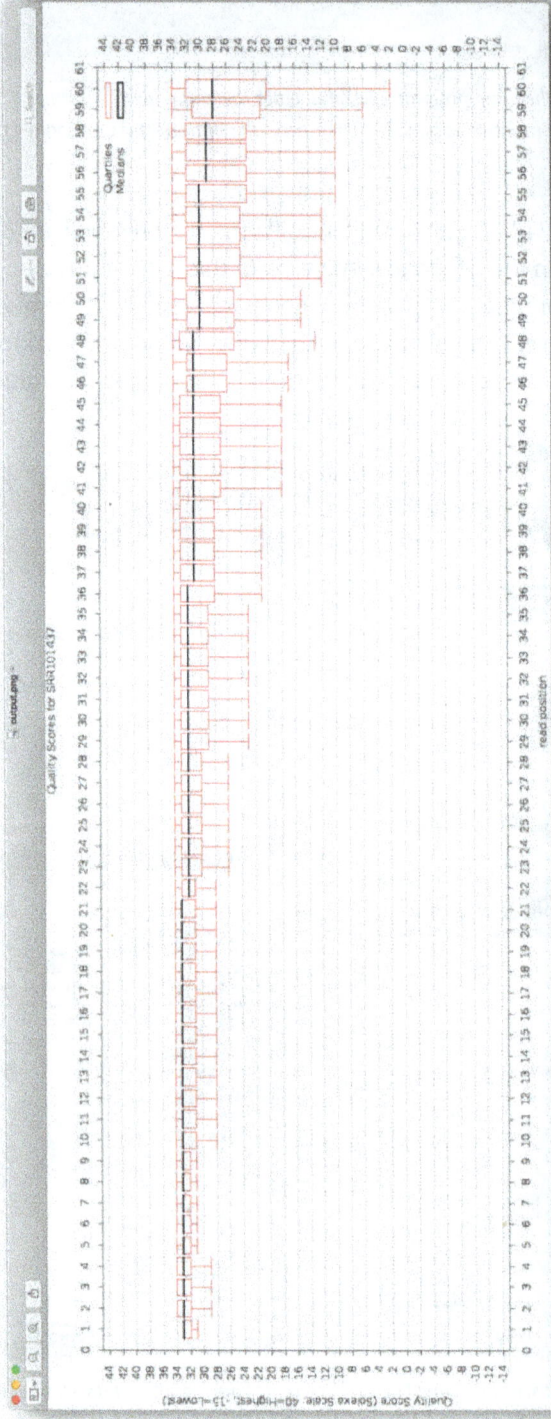

Figure 3.19 *Upon download of the PNG files back to your local PC, make use of any open software that can open either PNG or PostScript ps files, and you will see the results of the fastx_quality_stats analysis in the form of box plots, with helpful titles to guide you*

Through the above operation, I hope that readers can make use of applications on Linux command line in text form. You would have now experienced how to look at the help files and follow the instructions that guide you into selecting and entering the correct parameters in the correct syntax, let the program run and output the expected results for visualization on your local PC after file transfer.

We know that this series of operations may be difficult to follow, with having to read unfamiliar technical jargon. With mobile devices and the graphical user interface very much in vogue, this process may seem pretty archaic and inhuman. But this unfortunately is the current situation for bioinformatics researchers, especially if you want to use the latest freely accessible, open source software tools provided by other research institutions. The cost of foolproofing and dummy-proofing any software is high, and it takes a lot of time and resources to develop. Perhaps one day, tools like FASTX-Toolkit and others will attain the level of ease of usage like FastQC.

In comparison to the computational overheads of a Java virtual machine, which FastQC uses, the natively compiled FASTX-Toolkit binaries run at a performance level better than FastQC. When the original FASTQ input sequence file is very large, the benefits of the powerful FASTX-Toolkit application become very obvious. For the determined and intrepid, not deterred by the command line maneuverings, the FASTX-Toolkit package has applications with many features. Each program file name should be enough to give you hints as to what it does. Use the -h help option to understand how to use its parameters and options.

The process of the entire quality analysis generally follows what has been described above. However, do note that the approaches may differ between different research groups. The most important thing to note, is that high-throughput sequencing studies require stringent quality analysis, the main reason being that the accuracy of the sequence data is not necessarily ideal, due to the lack of maturity of next generation sequencing technologies. Based on practical experience, even with today's more mature technologies compared to the early days of shotgun sequencing, leading and trailing ends of each sequence fragment are generally of low quality. This is because each sequence fragment requires amplification using sequencing primers, which will add primer sequences at the front and rear ends. Moreover, later stages of the chemical reaction cycle tends to suffer from stability issues.

In practice, downstream sequence assembly procedures will require you to identify good quality regions and to trim off poor quality regions, based on the results of the quality analysis. This process involves trade-offs between too stringent a quality threshold resulting in too many sequences being trimmed off, and too low a threshold resulting in too many low quality sequences creeping into your final sequence and thus affecting the credibility of your work.

Typically, we can use the Phred score as a guide to ensure the quality of the sequence output, using the threshold of 30 (99.9% accuracy rate) as a rule of thumb. So, the procedure could involve first removing all sequences below this Phred Score threshold of 30 and then trim off any sequence beyond a fixed length of sequences. For example, take the average length of sequences above the Phred Score threshold, say the first 60 bases, and trim off everything else. Alternatively, this editing process can take place downstream for instance, during sequence alignment or other analytical steps.

CHAPTER 4

Sequence Alignment

4.1 The Purpose of Sequence Alignment

Of all the processes involved in high-throughput sequencing research, the technique of sequence alignment almost always appears at the beginning. Sequence comparison and alignment in bioinformatics research has been a very important issue from the beginning of the Human Genome Project from the late 1980s. Even from the earliest days of biocomputing to the bioinformatics as we know of today, researchers have sought to discover ever more effective ways of sequence alignment.

In bioinformatics, raw data from sequencing is converted into the simplest and most convenient abstraction, which is to represent the four different nucleotides of a DNA sequence as ATCG, and to save such data as strings in text files. The so-called sequence alignment in this regard, is to take two series of As, Ts, Cs and Gs in a sequence and compare them with each other and carry further analysis. In summary, the aim of sequence alignment in NGS is to place the sequence at the correct position of the entire genome.

Sequence alignment principles look simple enough to explain, but they are full of tough challenges, particularly with the advent of high-throughput sequencing technologies. This sequencing procedure is not necessarily of good quality. Scores are used to reference the quality of the regions of the sequence.

In addition, high-throughput automated sequencing tends to produce much shorter resulting sequences compared to traditional manual Sanger sequencing gels. So there is also the issue of sequence assembly, where it may be possible to locate these short sequence fragments at a lot of alternative positions on a reference sequence, so which one is the right location and which one is wrong? And during the generation of short sequences, randomly located errors also create many downstream problems? Given a mismatch, is it a sequencing error, or is it an evolutionary mutation, which occurs as an insertion in one or as a deletion in the other sequence in the comparison (indels). So how do we correct for sequencing errors and artefacts, as opposed to this naturally biological phenomenon called indels? This remains an open question which many are trying to resolve.

Because high-throughput sequencing generates so much data so quickly, both the sequence and its computation are big challenges. How fast can we speed up sequence alignment? Upgrading your computer's hardware is certainly one trivial solution, but improvement of the matching algorithm to solve the problem more efficiently is another challenge. In addition to speed, the question of sequence indels is also very important. Therefore, one of the holy grails of bioinformatics experts is to develop rapid, highly accurate, low-cost sequence alignment algorithms, preferably with specifications of hardware, which ordinary household PCs can fulfil. We will discuss the relatively few algorithms that have been varied upon in order to tackle the challenges.

What kind of information can we gather from the sequence alignment? There are a few simple applications. The first is the ability to infer the species from which the sequence comes from. Another is to infer function, based on sequence similarity. Similar sequences tend to originate from a common ancestor, and gene function is very likely to be preserved. So if a sequence is very similar to another sequence with known gene function, then this sequence is likely to possess the same function. Given a set of shot-gun sequence fragments from a species, if we already have the genomic sequence of the same or a closely related species, we can align these fragments onto the genomic sequence, also called the reference genome, and assemble the fragments therefrom into a complete genome. Sequence alignment is also used in disease research. From the genomic sequence of the patient, compared with that of a normal person, if we can find the difference, we may be able to derive a correlation between such differences or mutations, and may be able to identify mutations that cause genetic disease. For example, there is the well-known occurrence of a single nucleotide mutation in many positions within the different chromosomes of the human genome. These differences among different human individuals are called Single Nucleotide Polymorphisms (SNPs). If we can determine whether a single SNP or a specific set of SNPs correlate with a disease, genetic or otherwise, we may be able to better predict a disease condition and further diagnose the disease and initiate early or even possibly preventive treatment.

Of course the detailed process is not easy. Over millions of years of long evolutionary periods, the genome has already gone through many naturally occurring mutations, many of which confer evolutionary advantage and hence are fixed into the genome, while others are random mutations, which cause variations in the genomes, without serious impact to health. So there will be mutations found in the human genome, but whether a mutation is related to a particular disease, requires strong evidence in the form of large datasets in order to generate statistically significant correlations, and for subsequent biomedical experimentation to ultimately determine conclusively the causative link between mutation and disease.

Sequence assembly

Regarding the complex problem of the assembly of a reference genomic sequence, there are specialist bioinformatics scientists carrying out active research into sequence assembly methods. Previously, we have mentioned that with the existing reference sequences for sequence alignment, the process of sequence assembly of shot-gun sequence fragments, called Mapping Assembly, is a relatively simple approach, which involves positioning fragments onto the reference genome. If there is no reference sequence, genome assembly can be a big challenge. The assembly of a new genome sequence for the first time, or in other words, assembly without a reference genome (or *de novo* assembly) has been well-studied. One big problem is that if have insufficient overlaps of short shot-gun sequence fragments (also known as insufficient coverage), and if there are a lot of repetitive sequences, it is not easy to complete the genome sequence. There are numerous research papers (which can be searched for on websites such as PubMed or Google Scholar) on these issues if the reader is interested.

4.2 Selection of the Sequence Alignment Tools

In the context of Next Generation Sequencing, a simple approach for sequence alignment to a computer scientist or programmer might be string matching, which is to take a series of ATCGs in a string and compare it with the other. But such comparison in many cases for biological sequences like sequences in the genome, does not make much sense. As mentioned before, the sequence itself has intrinsically many issues from quality and accuracy to repetitive and short-read sequences, which the process of matching needs to consider. So there are a variety of sequence alignment algorithms that have emerged to manage this problem. Whether it is a commercial software or an open source software, depending on the techniques used, each software has its relative advantages and disadvantages. It also depends on the user requirements. One very important thing to consider is, especially in high-throughput sequencing era, is to focus on "short sequences", i.e. short reads comparison tools.

First generation sequence alignment, comparison and search tools such as those using Smith-Waterman local alignment algorithm, Needleman-Wunsch global alignment tools and of course, the famous and ubiquitous BLAST (Basic Local Alignment Search Tool), may not necessarily meet the needs of short sequences. With the development of short-read sequencing techniques in the past decades, short sequences may be in the range of 30 to 50 bases. Owing to the rapid developments with new high-throughput sequencing technology in recent years, the length of each sequence fragment may exceed 100 bases or more. So with how short is short being constantly re-defined, one should be aware of using the appropriate existing sequence alignment tool suitable for your sequencing platform.

Early sequence alignment tools are mostly based on the Smith-Waterman local alignment algorithm. Its main operational concept is to create an alignment matrix of the two sequences, and to use a scoring matrix to compute an alignment score of a comparison of two long sequences. This alignment score also takes into consideration matches and mismatches, as well as gap creation and gap extension penalties. Various alternative alignment paths are scored to determine the optimal alignment.

Although this method, also known as a dynamic programming approach, is accurate and will eventually produce an optimal alignment that maximizes matches, and minimizes gap creation and gap lengths for a given set of parameters, it is very computationally expensive, requiring much resources. One alternative work-around involves heuristics, and forms the basis of the ubiquitously used Basic Local Alignment Search Tool (BLAST), which has been particularly useful for sequence search of a sequence database. Most biologists and all bioinformatics researchers would have encountered it. However, in tackling the Smith-Waterman algorithm head-on, especially on the PC, enhancements to matrix calculations has seen rapid development in logical processor design. Multimedia Extensions (MMX) and Streaming SIMD (Single-Instruction Multiple-Data) Extensions (SSE) have been introduced into Intel processors. This has enabled many to implement a variety of platforms based on such processor enhancements to accelerate the Smith-Waterman algorithm by parallelizing at the instruction level, and taking advantage of next-generation support SSE instruction set processors, including that of Sony PlayStation 3 with its powerful floating-point operation functions. Others have even successfully used Field Programmable Gate Arrays (FPGA) to accelerate the computationally intensive S-W algorithm. More recently, using

CUDA (which will be discussed in a subsequent chapter), relying on CPU and GPU SIMD instructions to carry out concurrent CPU and GPU computations, algorithms have been developed to speed up S-W computations through employing SSE-based vector execution units as accelerators and GPU SIMD parallelization.

The Smith-Waterman algorithm, used in long sequences that come from the traditional Sanger sequencing method, allows for precise comparison and benefits significantly from computational acceleration, but for NGS short-read sequences, the benefit is correspondingly less, and so in view of the complexity of short-read assembly, the Smith-Waterman algorithm is not ideal.

As such, newer bioinformatics applications have mostly relied on the Burrows Wheeler Transformation algorithms. There are at least three popular ones: Bowtie, BWA and SOAP. Each has its particular characteristic. For instance, Bowtie emphasizes on speed, but the result is not necessarily accurate.

Bowtie itself incorporates the Quality Score and results in better alignment results, but whenever it encounters best results with two identical top scores, it randomly selects one. So hopefully, the other, which is not being selected, is not the correct result.

The concept of BWA is roughly the same as Bowtie, but it will output a variety of potentially correct alignments, and is not limited to what is purported to be the best result. For some studies, this kind of output result may be more in line with the actual needs of the sequencing work, allowing researchers to analyze and judge on their own, which of the top alignments is best in their context.

SOAP, which is developed by China's BGI, was recently updated from version SOAP2 to the latest SOAP3 (which features graphics card support, to be described in greater detail in a later section on the advantages of graphics card applications), emphasizes a new improved structure for the indexing system, and results in improved structure and in a more rapid yet accurate support for longer short-read sequences.

How do we assess bioinformatics tools? If you want to evaluate the performance, whether of speed or accuracy, we would use a "fake unknown" sequence dataset or a "publicly known genomic sequence" as a reference "control group" sequence for validation.

Burrows Wheeler

The Burrows-Wheeler Transformation (BWT) is the commonly used algorithm of several recently developed sequence alignment tools. Its most important feature is to utilize a compressible index, so that the capacity of the index becomes very small and can be loaded during the run-time execution of the application, entirely into the computer memory for direct sequence comparison without having to read from the data storage. Reading from memory *versus* reading from the hard disk is a lot faster. One point to note on the use of BWT applications, is that they need to build an index and this index cannot be shared, due to run-time modifications to particular annotations in the index records, which are specific to BWT. So even though the sequence to be analyzed is the same, the index of a Burrows-Wheeler Alignment (BWA) comparison tool cannot be used as an index for a Bowtie application (which is already into a release version of Bowtie2, and similarly, while the BWT algorithm is also used in indexing, a Bowtie index is not compatible with a Bowtie2 index).

Similarly, in the assessment of sequence alignment tool, the usual practice is to take a reference sequence through a series of changes, where the user can input the error rate, the indel size and other user-defined parameters, to produce artificial test sequences with quality scores, sequence errors and artificially generated indels.

Then we take these various sets of fake sequences through the alignment process of each sequence alignment tool, and compare the results to the original reference sequence and thus verify the accuracy of that sequence alignment tool. Because the original reference sequence is used, the answers are known, the entire group of fake sequence theoretically should be able to be re-aligned back to the original sequence. By incorporating various error rates and artificially created indels into the test sequences, we can measure the ability of the sequence alignment tool to recover the original reference sequence. Of course, the best tool would be the one which can correctly position each short-read sequence (containing the artificially generated errors) back to the correct position on the reference sequence, even if the sequencing platform produces a high error rate and the occurrence of many indels.

From a cursory glance of 2011 journal articles, one can observe that the sequence alignment software applications currently used by many researchers for the high-throughput sequencing include Bowtie, BWA, MrFast, MrsFast, Novoalign and SOAP. Using the above method to assess their respective performance, the conclusion is that Bowtie has good performance, while Novoalign can handle large sequence indels with very high accuracy.

But for the other tools, their performance are not easy to articulate. For example, some of the strengths of sequence alignment software in the colorspace format used by SOLiD sequencing platform, cannot be compared fairly with software using data from the Illumina platform. BWA's poor performance is because it lists all matching results, and hence dragging down the overall performance. Although

SOAP cannot compare with Novoalign's ability to handle large indels so well, but in seven tests of indel data, in terms of the overall accuracy SOAP was the best.

That is because the SOAP sequence alignment strategy splits a short-read sequence into a three-part comparison (two ends and the middle segments), allowing two match fails. This segmentation procedure is pretty clever, because in most cases of high-throughput sequencing, the two ends have the highest error rates. We have already discussed this phenomenon in a previous chapter. If you need accuracy with speed, but do not want to spend money (because the higher accuracy application, Novoalign, is a paid commercial software), maybe SOAP or SOAP2 is a good choice for you.

As for the speed performance, it depends largely on the index. In one study, it is stated that the time taken for indexing is inversely proportional to the comparison time. This means that the more time spent indexing, the shorter time needed for sequence comparison. Software that does not create an index, rather directly jumps to the comparison step, will take longer time than others, which do. This concept is not difficult to understand: the finer the index, the more the reference sequences are further analyzed, the longer it takes, but the comparison will be much faster.

Like the yellow pages of a telephone book, if the index is very fine and complete, taking into account every stroke, phonetic, radical, character and word, catering to more than just one user's search, one can find the phone number very quickly. So as long as the index is built just once and re-used, we can skip the indexing time for frequently used reference sequences.

Table 4.1 *Currently used high-throughput sequencing tools, sequence alignment algorithms, URL and remarks*

Software	Algorithm	URL	Remarks
Bowtie	BWT	http://bowtie-bio.sourceforge.net/index.shtml	
Bowtie2	BWT	http://bowtie-bio.sourceforge.net/bowtie2/index.shtml	Index format switched to FM index
BWA	BWT	http://bio-bwa.sourceforge.net/	
mr(s)FAST	Seed-and-extend	http://mrfast.sourceforge.net/	
Novoalign	Hash	http://www.novocraft.com/	Commercial Software
SHRiMP	Smith-Waterman	http://compbio.cs.toronto.edu/shrimp	For colorspace format
SOAP2	BWT	http://soap.genomics.org.cn/soapaligner.html	Not open source

In summary, sequence alignment tools have their relative advantages and disadvantages. Depending on your needs and requirements, you will have to make your own choice as to which tools are most suitable. For example, some sequence alignment tools are targeted at smaller genomes, such as yeast, and if you use it for the much larger human genome, it may not be suitable.

The BWT encoding-decoding algorithm

There are many software, which currently use the BWT algorithm, with slight modifications and tweaks here and there, to achieve incredibly high throughput, high performance sequence alignment. By compressing indexes of large sequence datasets, more can be loaded into memory and hence achieve a speed-up compared to previously used hash tables. This section will explain some key principles of BWT: simple-to-implement transformation, sorting and reversibility.

As an example, let the BWT algorithm code the word ^BANANA@ (where the head and tail, are represented by ^ and @ respectively). First, we generate all rotations of the word, each time by transferring the last character to the beginning. So in this case, we can generate for eight characters of the word, eight different rotations. Next, all these permutations are sorted in alphabetical order. The last column after sorting, is BNN^AA@A. This string can be easily and more efficiently compressed.

So, in this case, consecutive two N can be expressed as 2N, and so too, A, giving the final result as B2N^2A@A. For this simple example, maybe it is not obvious that there is a storage advantage. But in the case of raw sequence data, only four kinds of characters ATCG are involved, and so the chances of similar characters contiguous with each other is higher, and for genomes, often, there will be large genomic fragments with contiguous bases. The use of the BWT coding compression algorithm will therefore result in greater efficiency.

Decoding the string BNN^AA@A is also very simple. First, sort the string BNN^AA@A and then combine it with the first column, producing a two character combination: BA, NA, NA, ^B, AN, AN, @^, A@. Sort again, and combine this with the first column again, to generate a three character combination (BAN, NAN, NA@, ^BA, ANA, ANA, @^B, A@^) and so on, until all eight characters are generated and sorted. The final sort (8th), will generate a combination that was originally encoded ^BANANA@.

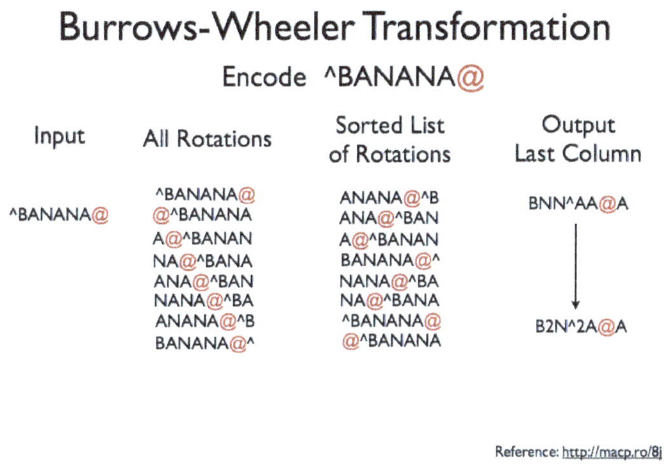

Burrows-Wheeler Transformation

Encode ^BANANA@

Input	All Rotations	Sorted List of Rotations	Output Last Column
^BANANA@	^BANANA@	ANANA@^B	BNN^AA@A
	@^BANANA	ANA@^BAN	
	A@^BANAN	A@^BANAN	
	NA@^BANA	BANANA@^	↓
	ANA@^BAN	NANA@^BA	
	NANA@^BA	NA@^BANA	
	ANANA@^B	^BANANA@	B2N^2A@A
	BANANA@^	@^BANANA	

Reference: http://macp.ro/8j

Figure 4.1 In encoding the string BANANA, the BWT algorithm after the rotations of the string, the last row is used as the new code, BNN^AA@A. You can also tie in with other compression methods to shorten the length of the encoded string

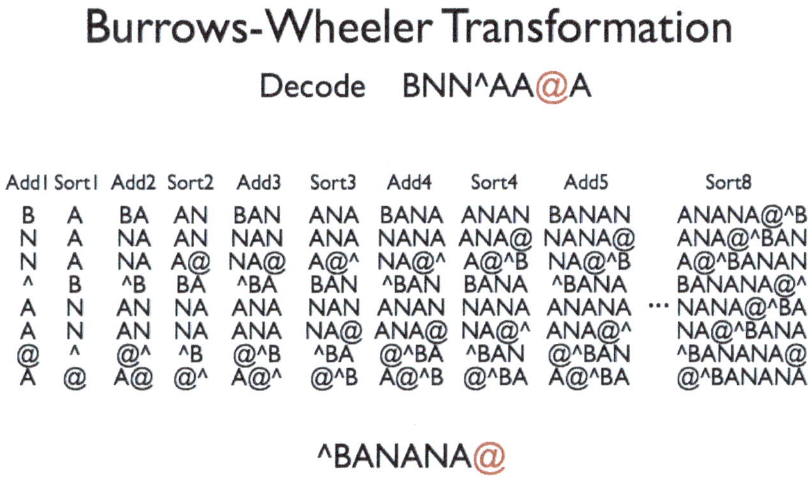

Burrows-Wheeler Transformation

Decode BNN^AA@A

Add1	Sort1	Add2	Sort2	Add3	Sort3	Add4	Sort4	Add5		Sort8
B	A	BA	AN	BAN	ANA	BANA	ANAN	BANAN		ANANA@^B
N	A	NA	AN	NAN	ANA	NANA	ANA@	NANA@		ANA@^BAN
N	A	NA	A@	NA@	A@^	NA@^	A@^B	NA@^B		A@^BANAN
^	B	^B	BA	^BA	BAN	^BAN	BANA	^BANA		BANANA@^
A	N	AN	NA	ANA	NAN	ANAN	NANA	ANANA	···	NANA@^BA
A	N	AN	NA	ANA	NA@	ANA@	NA@^	ANA@^		NA@^BANA
@	^	@^	^B	@^B	^BA	@^BA	^BAN	@^BAN		^BANANA@
A	@	A@	@^	A@^	@^B	A@^B	@^BA	A@^BA		@^BANANA

^BANANA@

Figure 4.2 For decoding the BWT-encoded string, BNN^AA@A, sort the encoded string, and combine with the encoded string sequentially. Repeat this until entire length of the original encoded string is traversed. For example, BNN^AA@A has eight characters, so after eight such sort-and-combine steps, the original string can be recovered

4.3 Actual Operation of the Sequence Alignment

The concept of sequence alignment tools are generally are the same. It is to take a query sequence and compare it with the reference sequence (or reference genome). The only difference is the choice of an algorithm.

This section will demonstrate the step-by-step operation of the sequence alignment process. The tool of choice, personally speaking, is Bowtie. The reference genome is the Human Reference genome, while the short-read sequence fragments from the sequencing platforms come from the public sequence databases. Each of the following paragraphs will explain the necessary file preparation and mode of operation of the commands. Once the reader becomes familiar with the sequence alignment operations, the same concept applies to other sequence alignment tools.

Download and installation of Bowtie

Bowtie, the short-read sequence alignment tool, was developed by researchers from the Center for Bioinformatics and Computational Biology, Institute for Advanced Computer Studies, University of Maryland. It is fast and consumes little memory. The goal is to be able to use the 2.2 GB of memory typically found in a PC, to complete a short-read sequence alignment with the human genome. There has been much effort to develop short-read sequence alignment tools, which are faster and more accurate than Bowtie, but at the moment, it seems to be in the lead. It is ultrafast, highly memory-efficient, and can align large sets of short DNA sequences to large genomes. For example, 35-base-pair reads can be aligned to the human genome at the rate of 25 million reads per hour on a typical workstation. The genome is indexed with a Burrows-Wheeler algorithm and this keeps the index memory footprint small. Note that it is not a general purpose alignment tool like BLAST or MUMmer, and cannot at this moment cope with long reads beyond 1 kb, or gapped alignments. Bowtie can be found on the SourceForge website under open source distribution, which you can freely download to use.

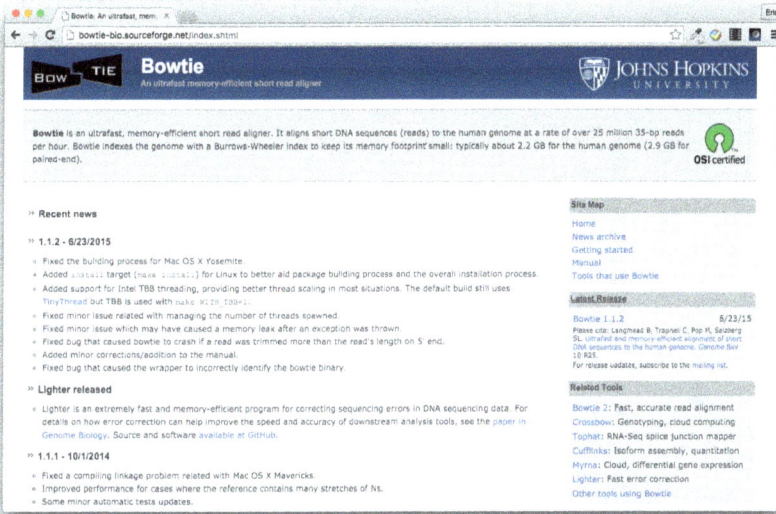

Figure 4.3 *Bowtie on SourceForge (http://bowtie-bio.sourceforge.net/index.shtml). Webpage providing complete user information*

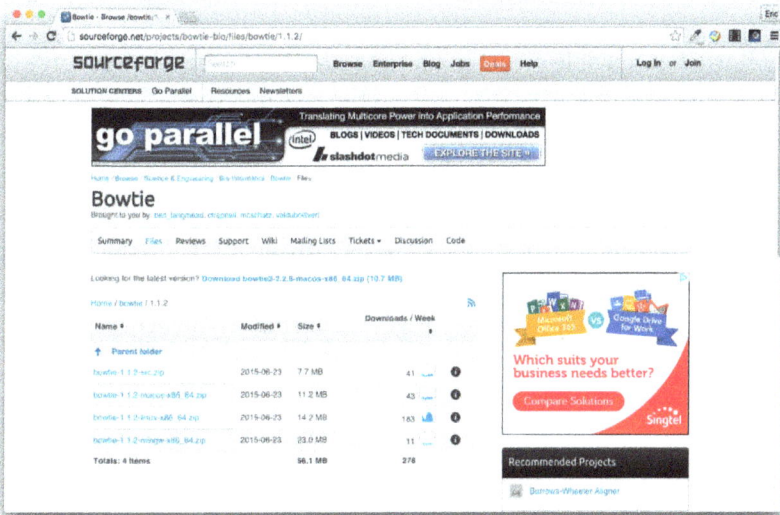

Figure 4.4 *Bowtie download webpage indicates that it is supported on several major platforms including macOS, Linux, Windows in 64-bit (x86_64) or 32-bit (i386) compatible binaries. You can also download the source code, unzip the folder and use the command line to compile the code yourself on your specific platform, if your platform and hardware architecture is not compatible with any of the above*

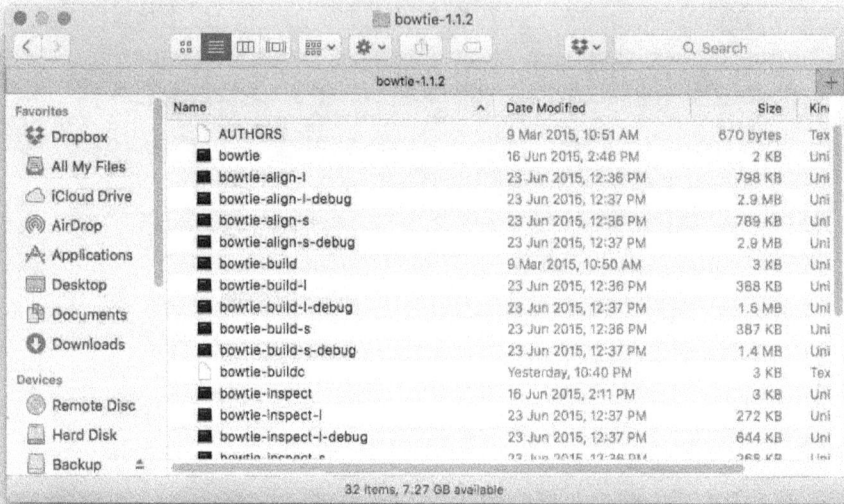

Figure 4.5 *Whether using the self-compiled or downloaded pre-compiled version, there are two very important programs that we need to use. One is bowtie, the other is bowtie-build. The former is the short-read sequence aligner main program, while the latter is used to build the index*

The Bowtie executable as the main program to use, can be used anywhere, so please remember the full directory path to the Bowtie program. Since you cannot use the original reference sequence for direct comparison, you need to build an index of the reference genome. At the Bowtie SourceForge web site, in fact, there are already a number of commonly used reference genome sequence pre-built indexes, which you can use immediately after downloading. If your genome reference sequence is not available, you will have to use the bowtie-build to generate a reference sequence index from your original reference sequence.

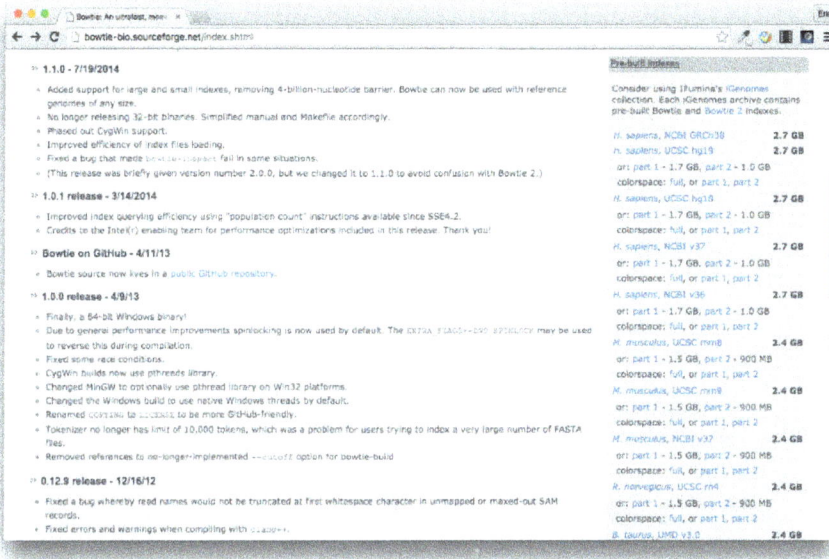

Figure 4.6 *If you scroll down the Bowtie home page, on the right panel, you can see a list of reference sequences. In addition to a complete reference sequence archive, there are split versions, part 1 and part 2. Similarly, there is another set, in colorspace format, which can be used for comparison. The colorspace version may be downloaded separately*

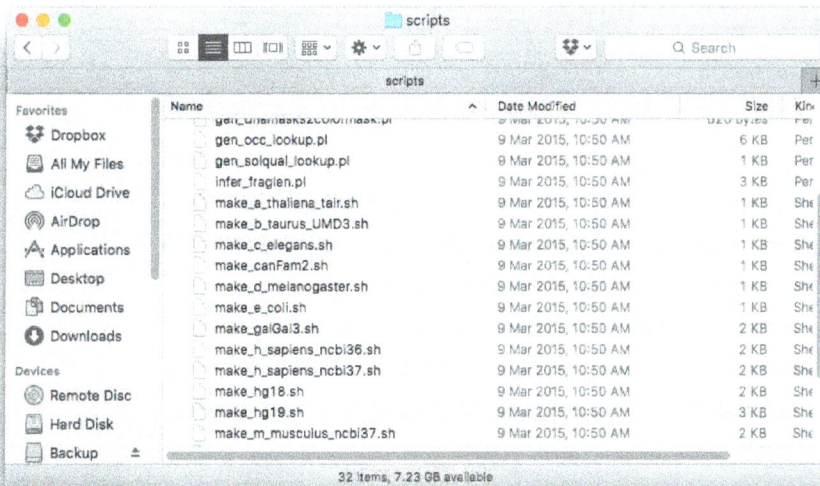

Figure 4.7 *In the downloaded Bowtie zip file after decompression, you may notice a scripts folder. You can see at the beginning of the series of files several UNIX shell script files (with .sh extension and "make" at the beginning of the file name) for a variety of species. These shell scripts are programs which, when executed at the command line, will automatically create/download an indexed reference sequence ready for use with bowtie*

```
● ● ●                      bowtie — -bash — 80×24
[Erics-Mac-mini:bowtie eric$ ls -alh
total 6004184
drwxr-xr-x  9 eric  staff   306B Nov 21 14:32 .
drwxr-xr-x  4 eric  staff   136B Nov 21 01:13 ..
-rw-r--r--  1 eric  staff   784M Nov 14  2009 hg19.1.ebwt
-rw-r--r--  1 eric  staff   341M Nov 14  2009 hg19.2.ebwt
-rw-r--r--  1 eric  staff   3.2K Nov 14  2009 hg19.3.ebwt
-rw-r--r--  1 eric  staff   682M Nov 14  2009 hg19.4.ebwt
-rw-r--r--  1 eric  staff   784M Nov 14  2009 hg19.rev.1.ebwt
-rw-r--r--  1 eric  staff   341M Nov 14  2009 hg19.rev.2.ebwt
-rw-r--r--  1 eric  staff   3.1K Nov 21 14:32 make_hg19.sh
Erics-Mac-mini:bowtie eric$ ▊
```

Figure 4.8 *To download pre-built human genome sequence indexes for bowtie alignment, for example, the H. sapiens, UCSC hg19 index, in the corresponding folder, you will see six main index related files: hg19.1.ebwt, hg19.2.ebwt, hg19.3.ebwt, hg19.4.ebwt, hg19.rev.1.ebwt, hg19.rev.2.ebwt. That make_hg19.sh, is the shell script that allows you to automatically download these six index files*

```
● ● ●     bowtie-1.1.2 — bowtie-build --wrapper basic-0 ../UCSChg19/hg19.fa hg19 — 80×...
[Erics-Mac-mini:bowtie-1.1.2 eric$ ./bowtie-build ../UCSChg19/hg19.fa hg19
Settings:
  Output files: "hg19.*.ebwt"
  Line rate: 6 (line is 64 bytes)
  Lines per side: 1 (side is 64 bytes)
  Offset rate: 5 (one in 32)
  FTable chars: 10
  Strings: unpacked
  Max bucket size: default
  Max bucket size, sqrt multiplier: default
  Max bucket size, len divisor: 4
  Difference-cover sample period: 1024
  Endianness: little
  Actual local endianness: little
  Sanity checking: disabled
  Assertions: disabled
  Random seed: 0
  Sizeofs: void*:8, int:4, long:8, size_t:8
Input files DNA, FASTA:
  ../UCSChg19/hg19.fa
Reading reference sizes
▊
```

Figure 4.9 *If you want to manually create a reference sequence Bowtie index, you have to run the bowtie-build. The command includes a bowtie-build reference sequence archive path and the path name to the hg19.fa file, my reference sequence is stored in the current folder (bowtie download folder) above the base folder of the UCSChg19 folder*

```
● ● ●                    bowtie-1.1.2 — -bash — 80×24
    offMask: 0xffffffe0
    isaRate: -1
    isaMask: 0xffffffff
    ftabChars: 10
    eftabLen: 20
    eftabSz: 80
    ftabLen: 1048577
    ftabSz: 4194308
    offsLen: 90540952
    offsSz: 362163808
    isaLen: 0
    isaSz: 0
    lineSz: 64
    sideSz: 64
    sideBwtSz: 56
    sideBwtLen: 224
    numSidePairs: 6467211
    numSides: 12934422
    numLines: 12934422
    ebwtTotLen: 827803008
    ebwtTotSz: 827803008
    reverse: 0
Total time for backward call to driver() for mirror index: 01:23:23
Erics-Mac-mini:bowtie-1.1.2 eric$
```

Figure 4.10 *Indexing time with the original size of the genome reference sequence and computer performance-related metrics as the bowtie-build command runs to completion. It took 1.47 minutes to create a 800 Mbyte file. Remember that unless you see the last line "*Total time for backward call to driver () for mirror index*" appear, and the emergence of the command input prompt, the index is not complete. One way to free the command prompt as mentioned in a previous chapter, is to push the program to background mode right at the beginning of the build run. This can be done using* Ctrl-Z *to suspend the program and bring up the command prompt, and then entering* bg *(background) to send the program into background. One useful way to run things in the background immediately is to use append & at the end of a command line command. In addition, you would generally want to view the output of the command. So do consider adding the* >& file.log *to the end of the command to pipe all logs to the file, file.log, which you can inspect from time to time, using the* tail -f file.log *command, to see if everything is going smoothly, and check if the last line is reached. Alternatively, you can create another login window*

To build the index into hg19, the instructions are:

```
./bowtie-build ../UCSChg19/hg19.fa hg19
```

After the execution begins indexing, a series of index files for hg19 will be created and because I did not specify a path for hg19, it will get built in the root directory of the current bowtie folder hg19.

Table 4.2 *List of Bowtie pre-compiled genome indexes*

Reference sequence name	Species
H. sapiens, UCSC hg18	Human
H. sapiens, UCSC hg19	Human
H. sapiens, NCBI v36	Human
H. sapiens, NCBI v37	Human
M. musculus, UCSC mm8	Mouse
M. musculus, UCSC mm9	Mouse
M. musculus, NCBI v37	Mouse
R. norvegicus, UCSC rn4	Rat
B. taurus, UMD v3.0	Cattle
C. familiaris, UCSC canFam2	Domestic dog
G. gallus, UCSC, galGal3	Domesticated chicken
D. melanogaster, Flybase, r5.22	Fruit fly
A. thaliana, TAIR, TAIR9	Plant
C. elegans, Wormbase, WS200	Worm
S. cerevisiae, CYGD	Yeast
E. coli, NCBI, st.536	Bacteria

Executing sequence alignment

In the execution of sequence alignment, Bowtie provides for many parameters, but does not fully document additional parameters. Using arguments specifying the files, containing the reference sequence index and the query short-read sequences, is sufficient to initiate sequence alignment.

In addition, specifying the output file storage format, for instance, -S, will convert the output file format into SAM. In addition, for multi-core computers, using the -p parameter to specify how many processor cores are requested to process the sequence alignment.

If you enter a number more than the actual available processor cores, Bowtie will automatically use up to the maximum number of cores. There are various alignments modes and one popular mode is the -n N alignment mode, which is the default, will determine which alignments are valid according to no more than N mismatches in the first L bases, where L (the seed) is a number 5 or greater (set with the -l parameter) on the high quality left end of the read. Bowtie allows from zero up to three mismatches. If it is not set, the default value is 2.

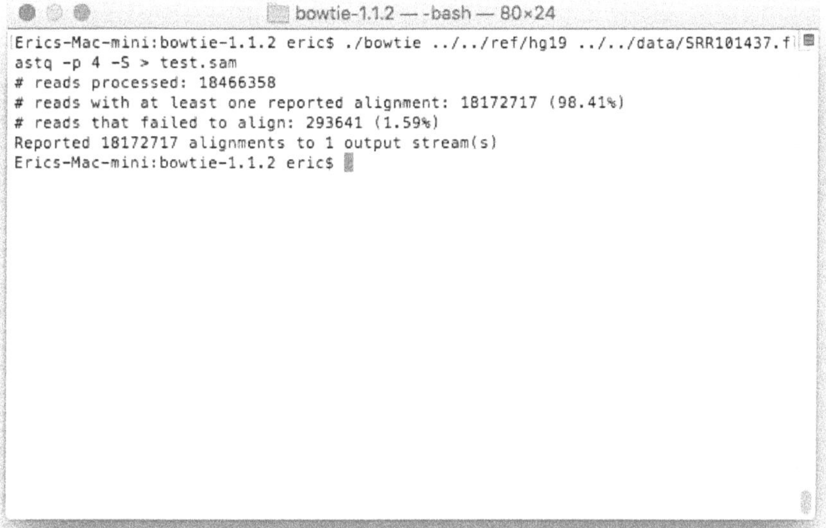

```
●  ◉  ●                    🔲 bowtie-1.1.2 — -bash — 80×24
Erics-Mac-mini:bowtie-1.1.2 eric$ ./bowtie ../../ref/hg19 ../../data/SRR101437.f 🔲
astq -p 4 -S > test.sam
# reads processed: 18466358
# reads with at least one reported alignment: 18172717 (98.41%)
# reads that failed to align: 293641 (1.59%)
Reported 18172717 alignments to 1 output stream(s)
Erics-Mac-mini:bowtie-1.1.2 eric$ ▌
```

Figure 4.11 *Suppose you want to compare the short-read sequences with the hg19 human reference sequence (the reader should adjust the various directory path to the specified files), and use the four processor cores and outputs into SAM format files, archive file name is test.sam. The complete command is:* bowtie ../../ref/hg19 ../../data/SRR101437.fastq -p 4 -S > test.sam

When sequence alignment is completed, in addition to a detailed comparison output to the specified output file, there are three aspects reported on:

1. Total number of processed sequences.
2. Number and proportion of success in sequence alignment.
3. Number and proportion of failures in sequence alignment.

From success ratio of these two numbers, we can get a preliminary assessment of similarity of the short-read sequences to the reference genome sequence.

After the match, the output reports that of the 18,466,358 short-read sequences (in SRR101437) being compared with the reference genome sequence, in this case hg19, 18,172,717 matches are found, accounting for 98.41%, the remaining 1.59% amounting to 293,641 unmatched short-read fragments.

```
● ○ ○               bowtie-1.1.2 — -bash — 80×24
Erics-Mac-mini:bowtie-1.1.2 eric$ ./bowtie ../../ref/hg19 ../../data/SRR101437.f
astq -p 4 -n 3 -S > test2.sam
# reads processed: 18466358
# reads with at least one reported alignment: 18162674 (98.36%)
# reads that failed to align: 303684 (1.64%)
Reported 18162674 alignments to 1 output stream(s)
Erics-Mac-mini:bowtie-1.1.2 eric$ ▌
```

Figure 4.12 bowtie hg19 SRR101437.fastq -p 4 -n 3 -S > test2.sam

With the same set of the above parameters, now adding a -n 3 parameter, which means that a sequence being compared against must possess less than the maximum of three bases mismatch. The output SAM file saved as test2.sam. Note the difference to the previous default -n 2. At a higher threshold, 1.64% not aligned

```
● ○ ○                 eric — top — 80×24
Processes: 353 total, 4 running, 10 stuck, 339 sleeping, 1769 threads  21:08:18
Load Avg: 4.28, 4.46, 3.30  CPU usage: 77.77% user, 17.25% sys, 4.96% idle
SharedLibs: 181M resident, 11M data, 7484K linkedit.
MemRegions: 122045 total, 5065M resident, 88M private, 683M shared.
PhysMem: 14G used (2924M wired), 1838M unused.
VM: 1011G vsize, 529M framework vsize, 98038287(0) swapins, 99070491(0) swapouts
Networks: packets: 132638339/65G in, 160348289/100G out.
Disks: 10113685/586G read, 10741500/734G written.

PID    COMMAND      %CPU    TIME      #TH  #WQ #PORT MEM    PURG  CMPRS
13972  screencaptur 0.2     00:00.03 5    3   57    1896K  20K   0B
13971  top          5.4     00:01.52 1/1  0   23    5312K  0B    0B
13966  bash         0.0     00:00.01 1    0   17    800K   0B    0B
13965  login        0.0     00:00.52 3    1   29    1152K  0B    0B
13963  bowtie-align 290.9   02:07.07 5/4  0   17    2254M  0B    0B
13962  com.apple.au 0.0     00:00.12 2    1   34    1500K  0B    0B
13961  com.apple.au 0.0     00:00.02 3    2   25    968K   0B    0B
13956  mdworker     0.0     00:00.03 3    0   49    1564K  0B    0B
13955  mdworker     0.0     00:00.03 3    0   49    1572K  0B    0B
13948  cupsd        0.0     00:00.06 3    0   43    1804K  0B    0B
13947  mdworker     0.0     00:00.05 3    0   59    1940K  0B    0B
13946  QuickLookSat 0.0     00:00.30 2    0   46    4452K  128K  0B
13945  ocspd        0.0     00:00.02 3    0   34    1372K  0B    0B
13944  quicklookd   0.0     00:00.80 4    0   93    6052K  0B    0B▌
```

Figure 4.13 *To find out the number of cores used, you can open up another terminal screen as mentioned in the previous chapter, and use the top command to monitor system resource usage. You can see the row reporting the data on the bowtie command indicates 290% CPU. In theory, in a 4 core system, we can use up to 400% of processor resources. This method can thus be used to monitor the run-time progress of Bowtie as it performs the sequence alignment*

For more details about the full specifications of using Bowtie, please visit their website and look for the latest release.

4.4 Sequence Alignment Results File Conversion

In previous sections, the final output of the comparison result is a SAM file (Sequence Alignment/Map), which is a kind of human-readable sequence comparison results shown in Tab-delimited format. Compared to other sequence alignment formats used to display comparison results, the biggest advantage is the use of few characters and a more streamlined way to record the comparison results. (Anyway, eliminating the need for a place, you can use specification defines after which you can restore into other formats, some information that you cannot miss.) The SAM format has an additional advantage. It can be compressed into a binary BAM (Binary Alignment/Map) format for accelerated processing. The actual operation uses SAMtools, a set of utilities for processing short DNA sequence read alignments by Heng Li. Processing of the BAM files can be segmented, without having to decompress or extract the entire SAM file, thereby saving system resources. After subsequent sorting and indexing, the performance of the BAM file will be even better.

There are many downstream bioinformatics analysis tools (Downstream analysis), to accept the results of sequence alignment to BAM files are based, but also require to have an index file. So we need to tie SAMtools to this tool before processing Bowtie SAM output file.

Downloading SAMtools

SAMtools (http://www.htslib.org/) is an open source project on the SourceForge site. From the SAMtools project page, you can download the Samtools package (http://www.htslib.org/download/). SAMtools does not have pre-compiled platform specific versions, so after downloading the latest versions from Github or SourceForge samtools-1.8.tar.bz2 (or whatever version it is), you have to compile the software yourself using

To decompress the tarball
```
tar -jxvf samtools-1.8,.tar.bz2
cd samtools-1.8
```

To compile
```
./configure --prefix=/where/to/install
make
make install
```

To read the manual,
```
MANPATH=/where/to/install/share/man:$MANPATH
export MANPATH
man samtools
```

To make the version of samtools which you have just installed executable from any directory, modify the default path as follows.
```
PATH=/where/to/install/bin:$PATH
```

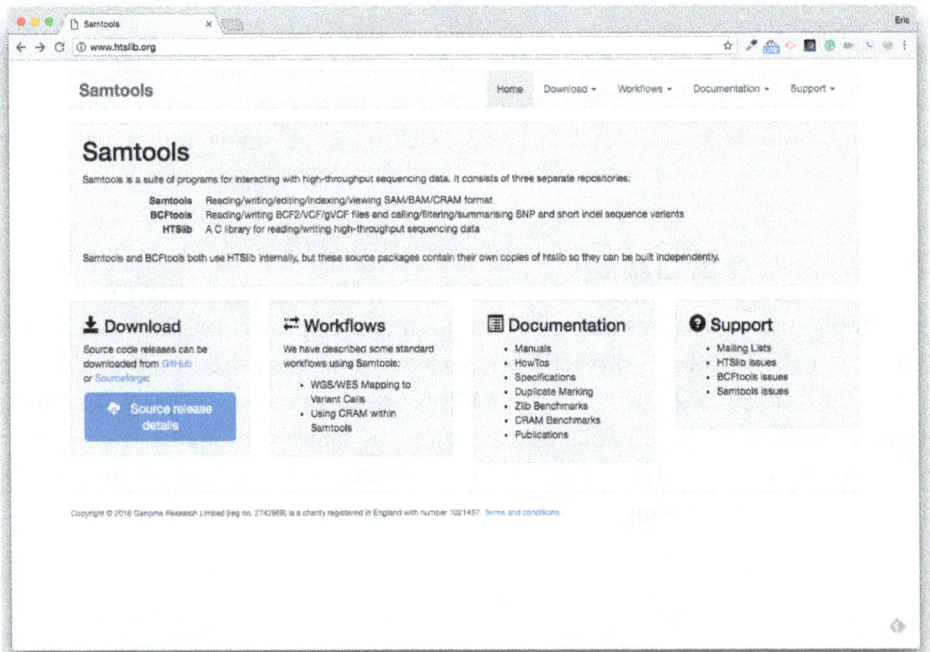

Figure 4.14   *Homepage of SAMtools on SourceForge — http://www.htslib.org. Click SourceForge link in Download section*

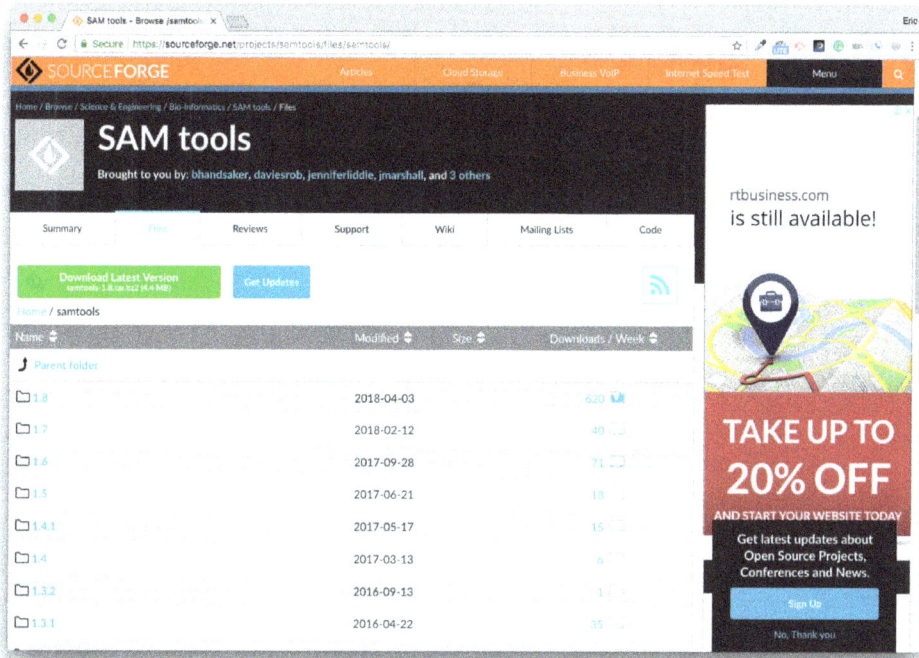

Figure 4.15 *In SAMtools SourceForge page, Files tab, there are many links to various versions of SAMtools. Click for the wanted version, or just click the green button to download the latest release*

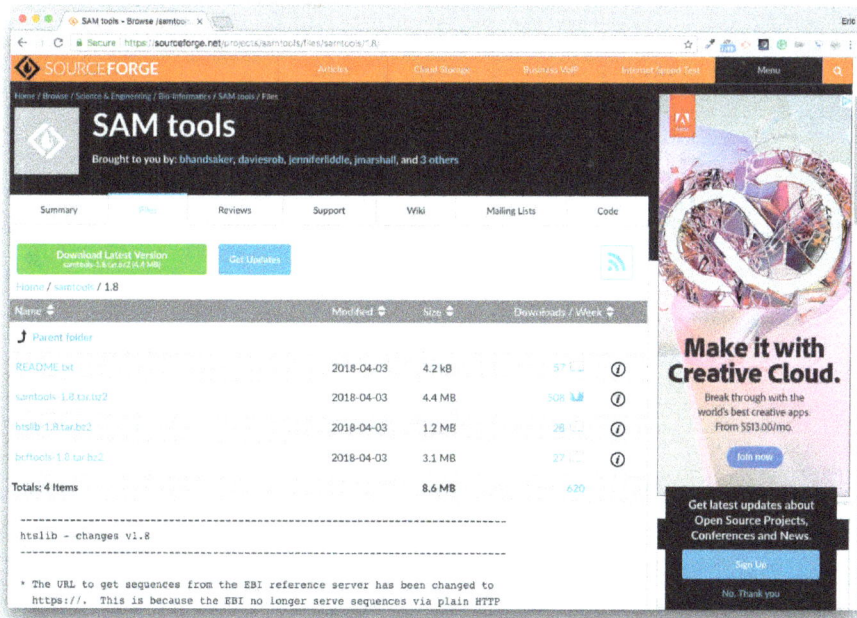

Figure 4.16 *Click on the Bzip2 compressed packed files (samtools-1.8.tar.bz2) to start the download*

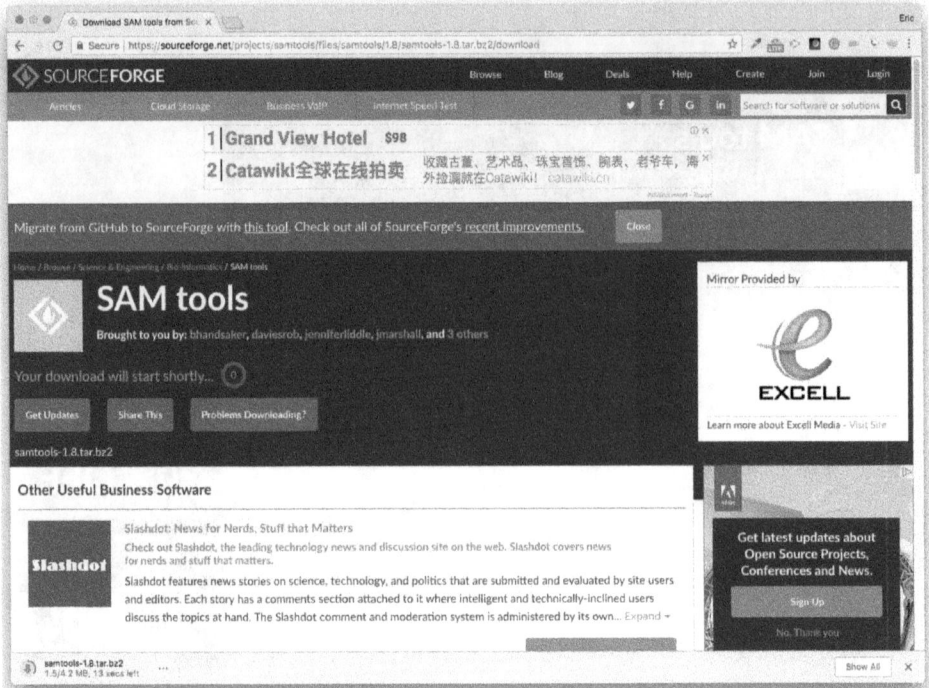

Figure 4.17 *Start to download the packed files*

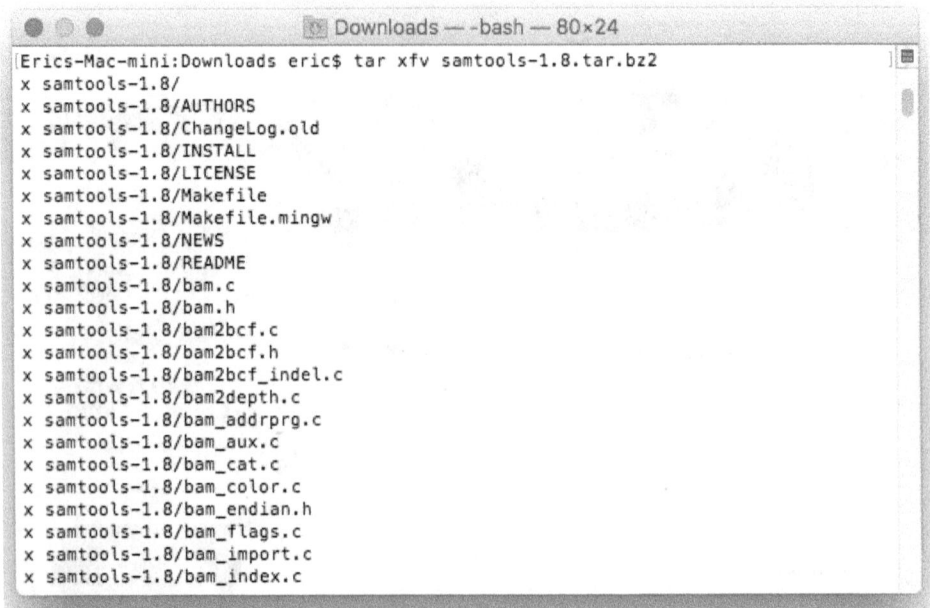

Figure 4.18 *After download is completed, use the command with the parameters* xvf *to untar and decompress. Complete command is* tar xfv samtools-1.8.tar.bz2, *which will extract all the files to a folder called samtools-1.8*

```
● ◎ ●                    samtools-1.8 — -bash — 80×24
Erics-Mac-mini:samtools-1.8 eric$ make
Makefile:110: config.mk: No such file or directory
config.mk:34: htslib-1.8/htslib_static.mk: No such file or directory
cd htslib-1.8 && /Applications/Xcode.app/Contents/Developer/usr/bin/make htslib.
pc.tmp
sed -e '/^static_libs=/s/@static_LIBS@/-lz -lm -lbz2 -llzma/;s#@[^-][^@]*@##g' h
tslib.pc.in > htslib.pc.tmp
cd htslib-1.8 && /Applications/Xcode.app/Contents/Developer/usr/bin/make htslib_
static.mk
sed -n '/^static_libs=/s/[^=]*=/HTSLIB_static_LIBS = /p;/^static_ldflags=/s/[^=]
*=/HTSLIB_static_LDFLAGS = /p' htslib.pc.tmp > htslib_static.mk
echo '/* Basic config.h generated by Makefile */' > config.h
echo '#define HAVE_CURSES' >> config.h
echo '#define HAVE_CURSES_H' >> config.h
gcc -g -Wall -O2 -I. -Ihtslib-1.8 -I./lz4  -c -o bam_index.o bam_index.c
gcc -g -Wall -O2 -I. -Ihtslib-1.8 -I./lz4  -c -o bam_plcmd.o bam_plcmd.c
gcc -g -Wall -O2 -I. -Ihtslib-1.8 -I./lz4  -c -o sam_view.o sam_view.c
gcc -g -Wall -O2 -I. -Ihtslib-1.8 -I./lz4  -c -o bam_cat.o bam_cat.c
gcc -g -Wall -O2 -I. -Ihtslib-1.8 -I./lz4  -c -o bam_md.o bam_md.c
gcc -g -Wall -O2 -I. -Ihtslib-1.8 -I./lz4  -c -o bam_reheader.o bam_reheader.c
gcc -g -Wall -O2 -I. -Ihtslib-1.8 -I./lz4  -c -o bam_sort.o bam_sort.c
gcc -g -Wall -O2 -I. -Ihtslib-1.8 -I./lz4  -c -o bedidx.o bedidx.c
gcc -g -Wall -O2 -I. -Ihtslib-1.8 -I./lz4  -c -o bam_rmdup.o bam_rmdup.c
gcc -g -Wall -O2 -I. -Ihtslib-1.8 -I./lz4  -c -o bam_rmdupse.o bam_rmdupse.c
```

Figure 4.19 Use the cd command to switch to samtools-1.8 folder, enter the make command to compile the source code

```
● ◎ ●                    samtools-1.8 — -bash — 80×24
Erics-Mac-mini:samtools-1.8 eric$ ./samtools

Program: samtools (Tools for alignments in the SAM format)
Version: 1.8 (using htslib 1.8)

Usage:   samtools <command> [options]

Commands:
  -- Indexing
     dict           create a sequence dictionary file
     faidx          index/extract FASTA
     index          index alignment

  -- Editing
     calmd          recalculate MD/NM tags and '=' bases
     fixmate        fix mate information
     reheader       replace BAM header
     targetcut      cut fosmid regions (for fosmid pool only)
     addreplacerg   adds or replaces RG tags
     markdup        mark duplicates

  -- File operations
     collate        shuffle and group alignments by name
     cat            concatenate BAMs
```

Figure 4.20 Once compiled, make sure you are in the samtools-1.8 folder, enter the command ./samtools, and you should see the dialogue window for a description of other performance parameters screen. This means that the main program SAMtools is now working

```
● ● ●                    ⬆ eric — ssh root@cctv.eric.lv — 80×24
0 upgraded, 1 newly installed, 0 to remove and 0 not upgraded.
Need to get 563 kB of archives.
After this operation, 1292 kB of additional disk space will be used.
Get:1 http://http.debian.net/debian/ jessie/main samtools amd64 0.1.19-1 [563 kB
]
Fetched 563 kB in 2s (233 kB/s)
Can't set locale; make sure $LC_* and $LANG are correct!
perl: warning: Setting locale failed.
perl: warning: Please check that your locale settings:
        LANGUAGE = (unset),
        LC_ALL = (unset),
        LC_CTYPE = "UTF-8",
        LANG = "en_US"
    are supported and installed on your system.
perl: warning: Falling back to a fallback locale ("en_US").
locale: Cannot set LC_CTYPE to default locale: No such file or directory
locale: Cannot set LC_ALL to default locale: No such file or directory
Selecting previously unselected package samtools.
(Reading database ... 44173 files and directories currently installed.)
Preparing to unpack .../samtools_0.1.19-1_amd64.deb ...
Unpacking samtools (0.1.19-1) ...
Processing triggers for man-db (2.7.0.2-5) ...
Setting up samtools (0.1.19-1) ...
root@cctv:~# ▊
```

Figure 4.21 *Using the Ubuntu repository to install SAMtools generally gets you an older version and not the latest release. Whether to use this simple way to install is up to the reader's discretion*

If you had adopted our recommendations earlier and installed Ubuntu Linux, then you can install from Ubuntu repositories directly as they will have SAMtools.

On your Ubuntu, enter

```
sudo apt-get install samtools
```

This command will prompt asking whether to install SAMtools or not. Enter Y to start installation. This process is equivalent to the earlier process of compilation using the UNIX make command. But note that the SAMtools version in the Ubuntu package library repository is up to version 0.1.19.

For interconversion between SAM and BAM and other functionalities, SAMtools provides a lot of parameters used in the SAM conversion to BAM, including the parameter called view. Add the -b parameter to create a BAM format output file.

The -S parameter indicates that the input file is a SAM file. So the two parameter combination becomes -bS, followed by the SAM file you want to convert (don't forget the path) and the UNIX redirection for output to save into the archive BAM file location and file name.

```
● ○ ●  ☆ Eric — ubuntu@Erics-Macbook-Pro: ~/samtools-1.2 — ssh ubuntu@10.12.2.15...
ubuntu@Erics-Macbook-Pro:~/samtools-1.2$ ./samtools view -bS ~/result/result.sam
  > ~/result/result.bam
ubuntu@Erics-Macbook-Pro:~/samtools-1.2$ ▋
```

Figure 4.22   *Suppose you want to convert a SAM file result.sam stored in the root directory under the sub-folder result, the path to the new archive folder is also the same, result, with the file name called result.bam, the BAM file. In SAMtools main program directory, the complete command to run is:*
`./samtools view -bS ~/result/result.sam > ~/result/result.bam`

```
● ○ ●  ☆ Eric — ubuntu@Erics-Macbook-Pro: ~/result — ssh ubuntu@10.12.2.152 — 80...
ubuntu@Erics-Macbook-Pro:~/result$ ls -lh
total 4.6G
-rw-rw-r-- 1 ubuntu ubuntu 1.3G Nov 19 08:21 result.bam
-rw-rw-r-- 1 ubuntu ubuntu 3.4G Nov 19 08:14 result.sam
ubuntu@Erics-Macbook-Pro:~/result$ ▋
```

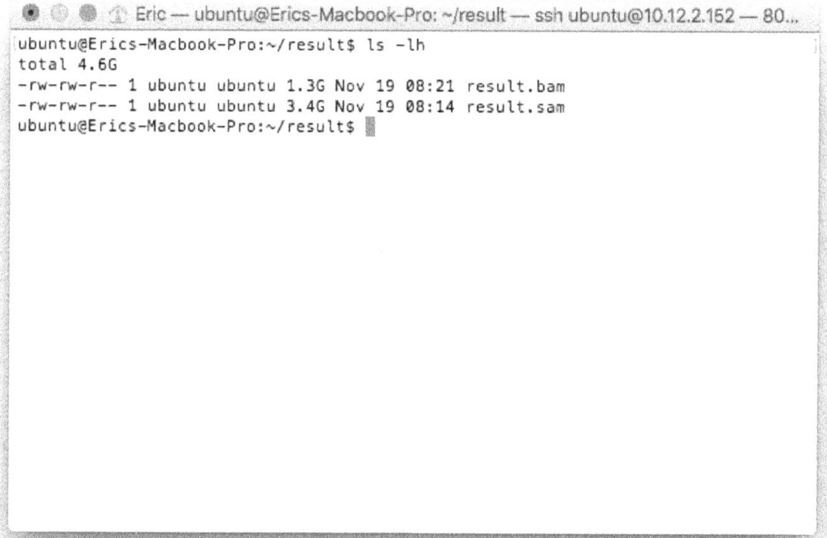

Figure 4.23   *Once the conversion is finished, switch to the root directory of result folder and as expected, there is a new file called result.bam with file size close to one-third of the SAM file. Evidently, conversion to BAM format saves disk space*

After sorting and indexing to generate BAM files, the task is only half done. Many BAM bioinformatics applications require BAM input files to be sorted and indexed. So we have to use the SAMtools main program with two other parameters to complete the job. The sequence must be sorted first; next, create its index. After the sorted BAM file and the corresponding index BAI file is generated, it is in a much more compact form which most bioinformatics tools can take, and generally more efficient during execution.

In the samtools-1.2 directory, execute the samtools main program, with the sort parameter rather the view parameters used previously. Follow that with the BAM file to be sorted, making sure the file path is correctly inputed. The complete command to execute is:

```
./samtools sort ~/result/result.bam
~/result/result.sort
```

Note the new file is named result.sort and appended with a .bam extension. This is because new files generated by SAMtools have .bam automatically appended, e.g., result.sort.bam. So if you enter an input file result.sort.bam, the output file becomes result.sort.bam.bam!

Figure4.24 *When you use SAMtools sort, a new output file has a .bam extension automatically added*

Figure 4.25   *The command for indexing is even easier. Just change the parameters of the main program to index, followed by the sorted BAM file with full path. The command to execute following the above example is:* ./samtools index ~/result/result.sort.bam

Figure 4.26   *Checking the contents of this folder, result, in addition to the original result.sam and the converted file result.bam and the sorted result.sort.bam, there is a new output file, result.sort.bam.bai. Please do not arbitrarily change the base file name. Otherwise, bioinformatics tools will not be able to follow through the series of converted, sorted and indexed BAM files which are related to each other*

4.5 Using the Genome Browser

After sequence alignment, we should inspect the outcome and view the reference sequence in conjunction with the short-read sequences. For this, we need to use a program called Genome Browser. Currently, many commercial and non-profit organizations are actively developing online and standalone versions. The main functionalities are the same, that is, to view at the finest resolution how the short-reads are aligned, and combined into longer sequences, and their corresponding genes in the reference genome sequences.

For example, if we are comparing the sequence of a patient suffering from a rare genetic disease, some locations in the genome which we cannot compare might be sites of mutation. Through the Genome Browser, the sequence alignment of the patient's genome with the "original" reference sequence plus the gene annotation data (gene annotation refers to the information describing the gene and its function, chromosomal location, etc., which are inserted after scientific research and findings have provided sufficient data and evidence to support the annotation). Together with the alignment data, we can observe if other similar patients with similar conditions possess common gene mutations. The results of the bioinformatics computation initially may be in plain text, but using the Genome Browser, we will have an informative graphical display to increase the effectiveness of our study.

There are many genome browsers currently available, but we prefer a standalone version, preferably not based on Java, because low memory situations occur frequently, especially when relatively large sequences or their alignment results are loaded. We also need to take into consideration the frequency of maintenance, whether continuously updated or not, and the ability to cope with the latest version of the operating system. Given these issues, Genome Workbench, developed by the U.S. National Center for Biotechnology Information (NCBI), is highly recommended. Examples of data, which it can utilize, includes sorted and indexed BAM files previously discussed.

Table 4.3   *List of existing commonly used genome browsers*

Name	URL	Remarks
Integrative Genomic Viewer, IGV	http://software.broadinstitute.org/software/igv/	
UCSC Genome Browser	https://genome.ucsc.edu	Online version
Ensembl	http://www.ensembl.org/index.html	
GenoViz	https://sourceforge.net/projects/genoviz/	Also known as IGB

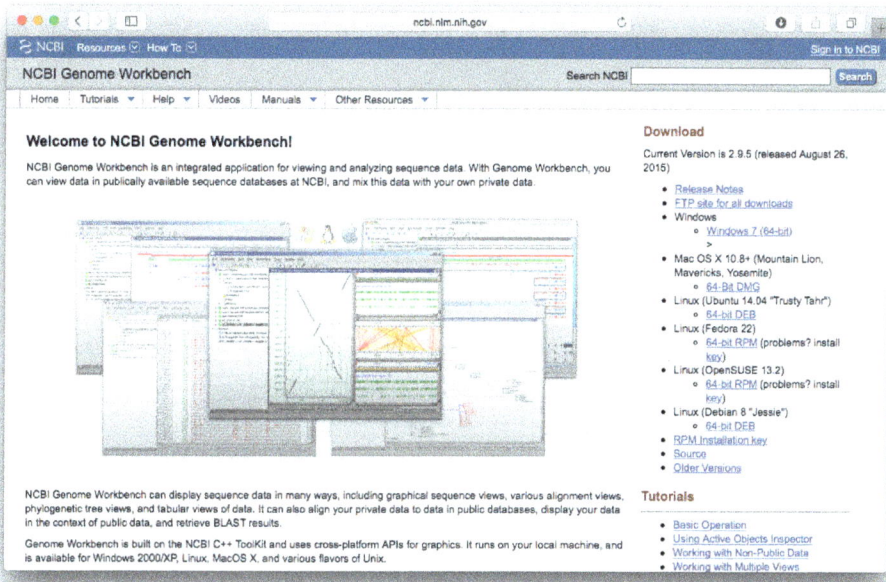

Figure 4.27 The NCBI Genome Workbench website (www.ncbi.nlm.nih.gov/tools/gbench), the right side of which has links to Genome Workbench compiled on various operating systems. Download the appropriate version. We use the Mac version of the platform in the subsequent demonstrations below. Genome Workbench operating on other platforms are similar

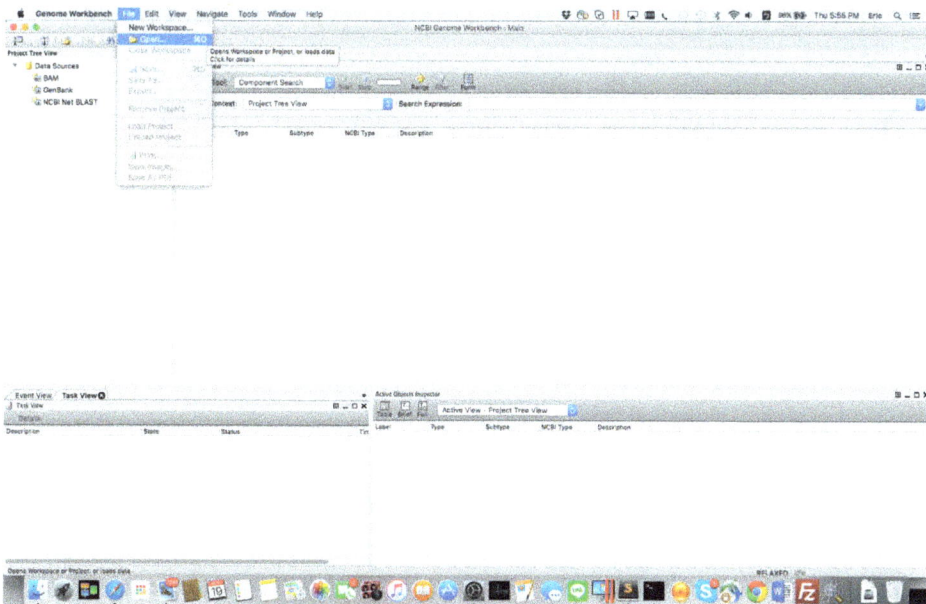

Figure 4.28 Once the network connection is working, the main program starts up. To load a BAM file, select from the dropdown menu, File, and choose Open to load the desired file

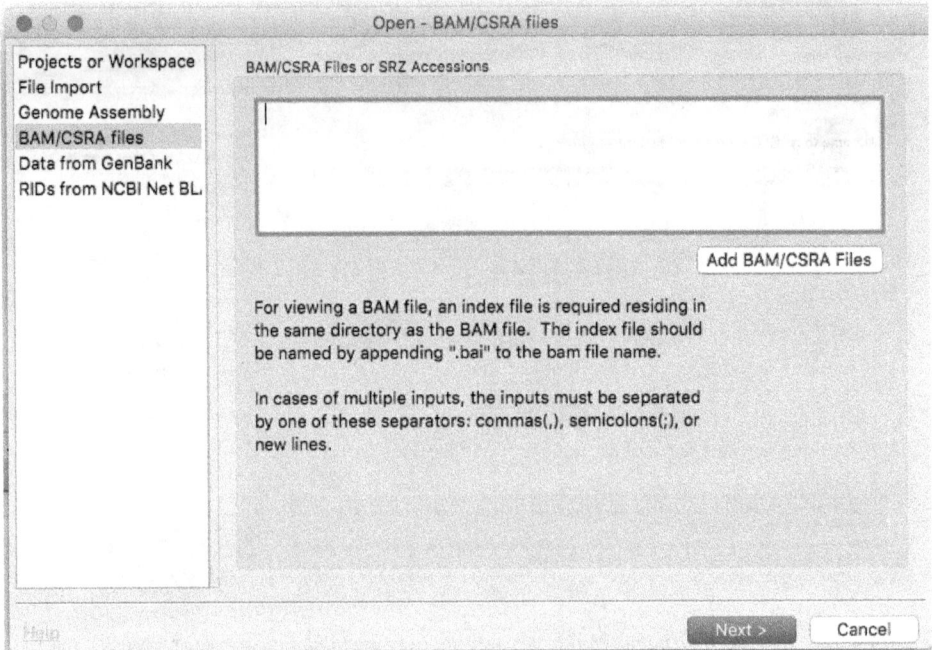

Figure 4.29 *Many file types are supported. From the left of the window, select BAM or CSRA files as the appropriate type of file you wish to load. Click the "*Add BAM/CSRA Files*" button to select your desired BAM file*

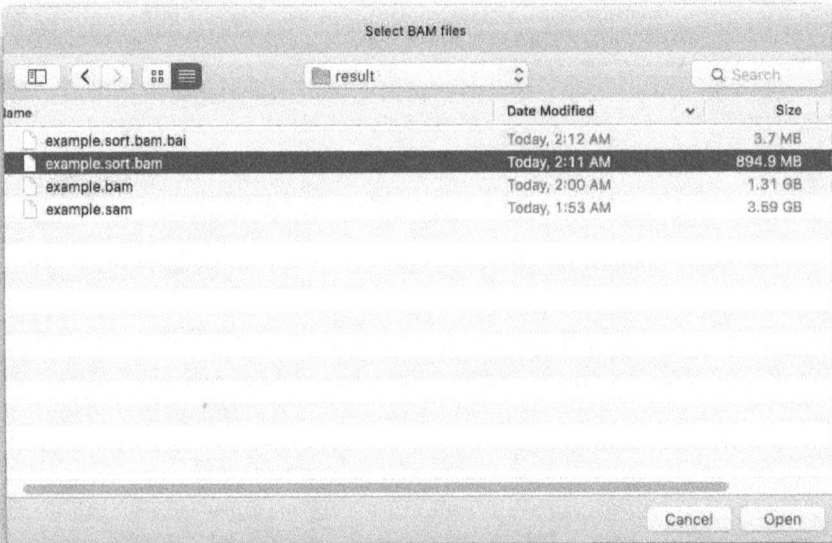

Figure 4.30 *From the file browser dialog window, select the BAM file which was created earlier. Before you can open the example.sort.bam, check that its corresponding index file, example.sort.bam.bai exists*

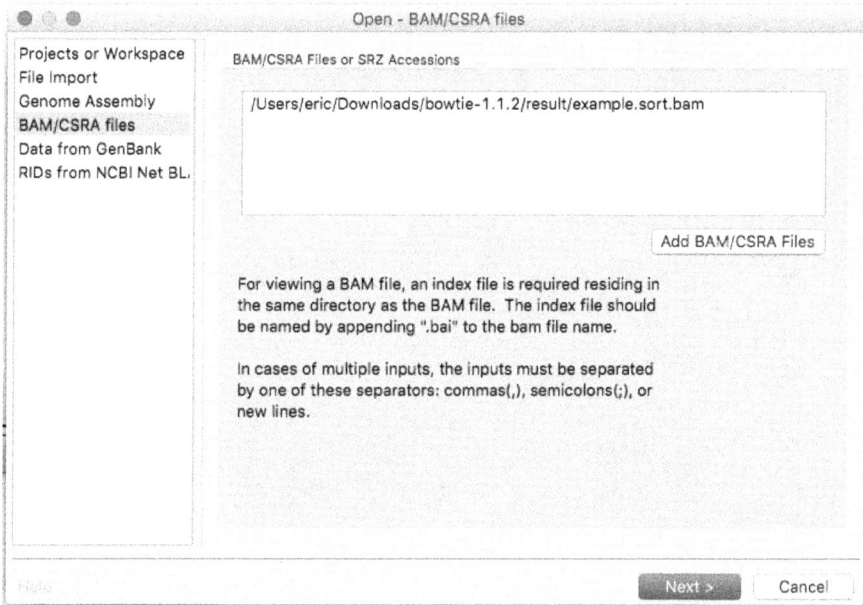

Figure 4.31 Check that the new window that pops up is pointing to the correct full path of the desired file; then, press the Next button to continue

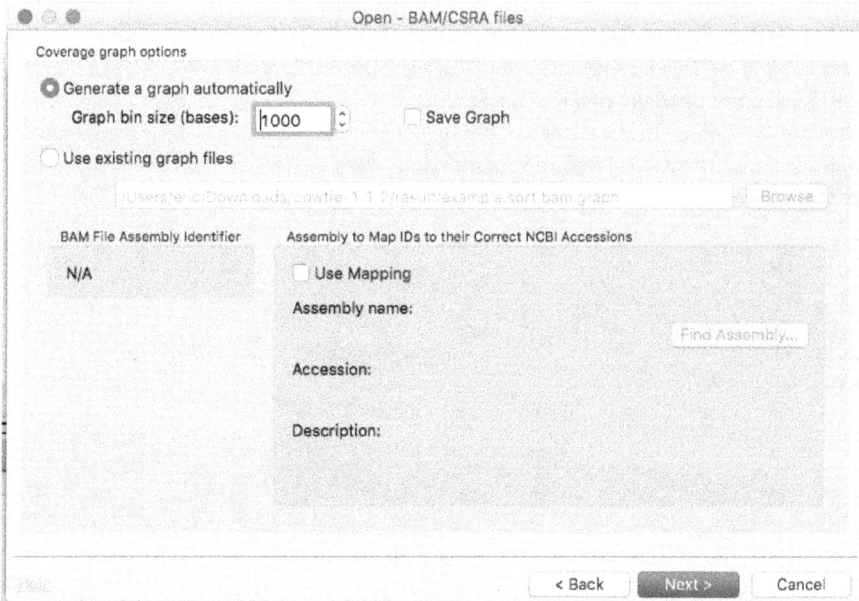

Figure 4.32 To display a figure showing the coverage (coverage refers to the number of short-read sequences providing data support to a particular section of the reference sequence), use the default "Generate a graph automatically". Please also check the option "Use Mapping", and press the "Find Assembly" button

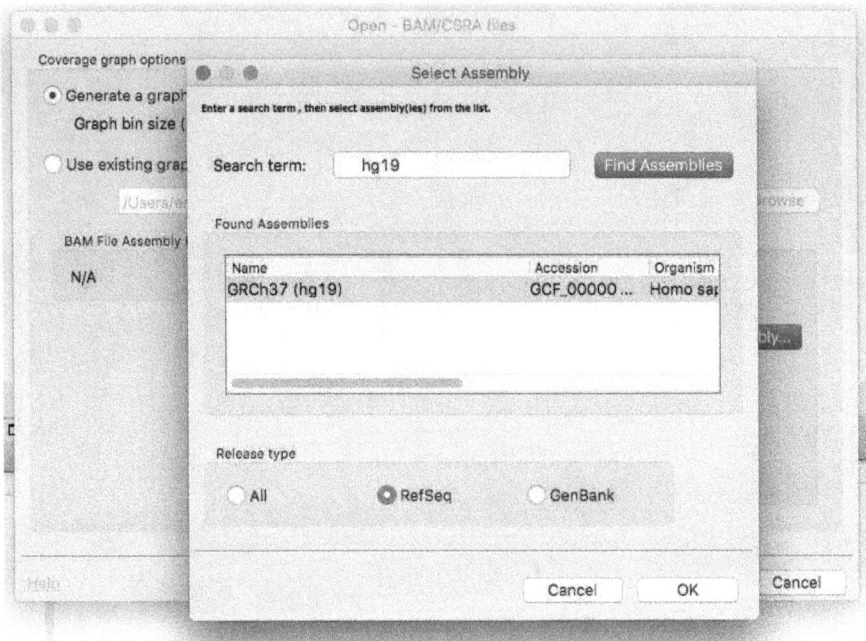

Figure 4.33 *In providing search keyword for your reference sequence, do note that a particular species usually has a public version of the reference sequence. Since our choice before the sequence alignment was to use a reference genome sequence, in this case, the hg19 human genome sequence, so please input h19. The result of the search using hg19 shows that there are two kinds of reference assemblies, RefSeq and GenBank. Since GenBank contains duplicate data, while RefSeq derives from GenBank detailed non-duplicate data, please choose* RefSeq. *But note that new data deposited with Genbank, may not yet be curated in the latest version of RefSeq. So in such cases, when we need the most recent data, we should choose the* GenBank *instead. Select the appropriate one and then press* OK *to continue*

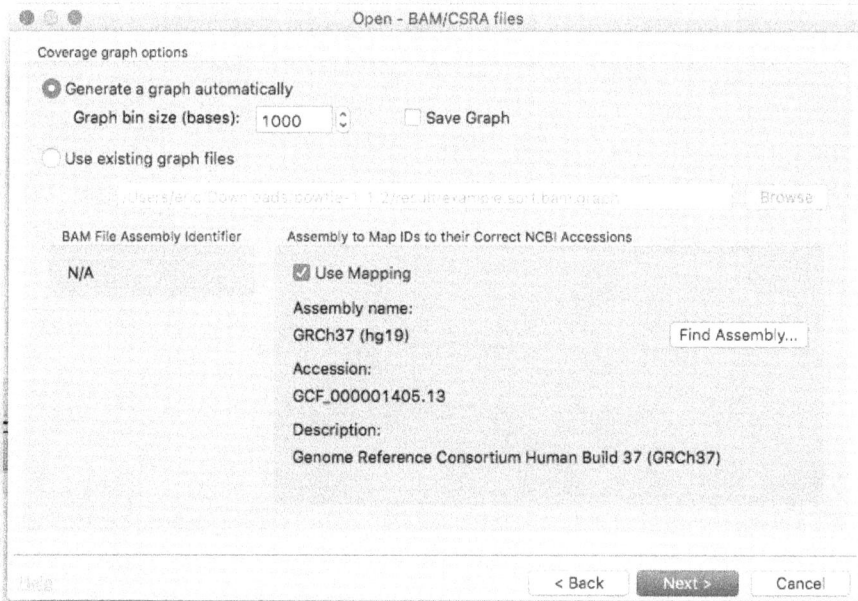

Figure 4.34 *Back to Coverage graph options dialog window, press* Next *to continue*

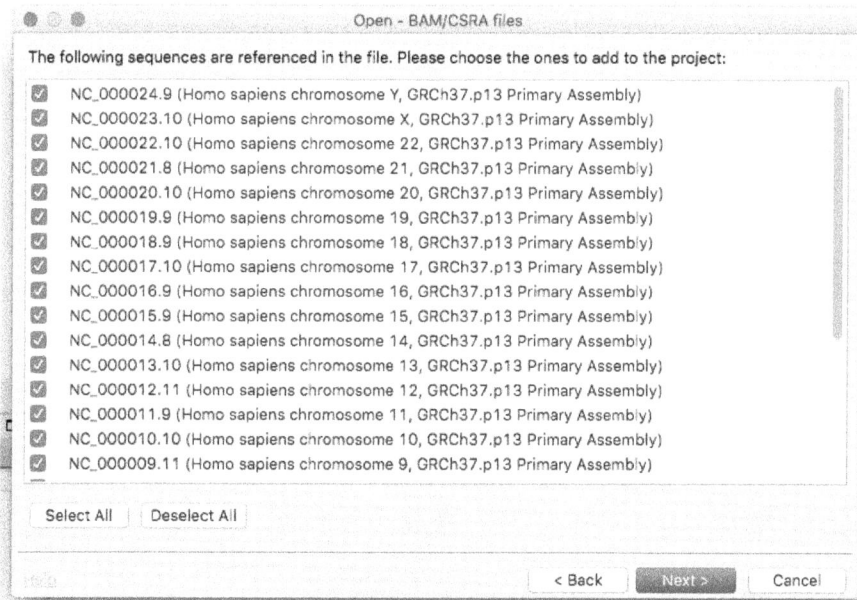

Figure 4.35 *Referenced sequences listed and clustered by Chromosome. Press* Next *button to load all of them*

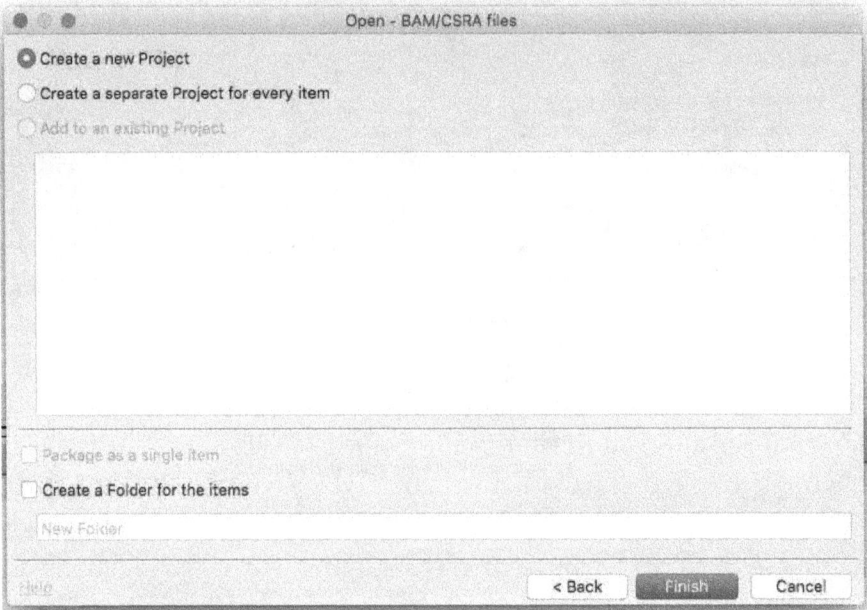

Figure 4.36   *Finally, select* create a new Project, *which presents the information in a new separate display. Press the* Finish *button to complete the process of loading the BAM file*

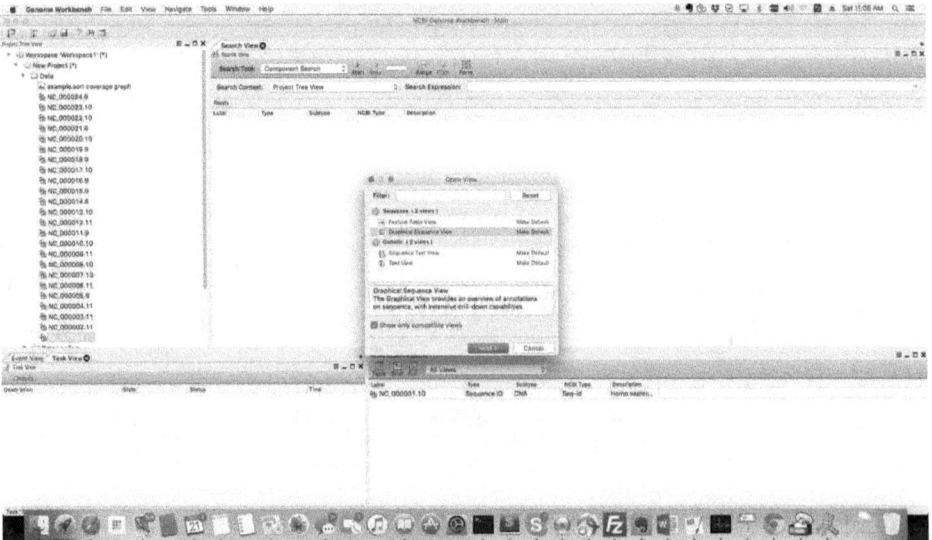

Figure 4.37   *After successfully loading the BAM file and generating a new project, as seen from the left pane of the main screen, start the graphical view by double-clicking example.sort coverage graph. You will be prompted with dialog window whereupon please select* Graphical View *therein and press* Next. *If readers have other requirements, you can choose to view the other items*

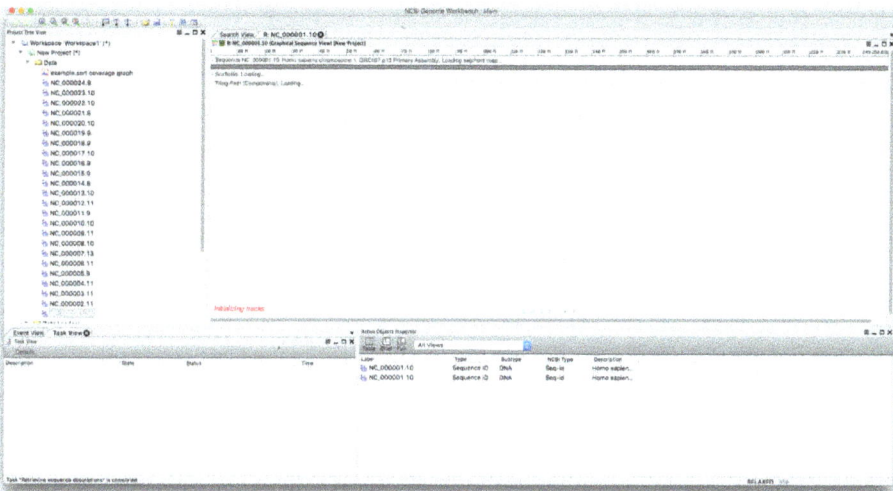

Figure 4.38 *Genome Workbench open new tab page, download all necessary data for data visualization from Internet, load the reference sequence and your alignment result. These procedures need some time to process*

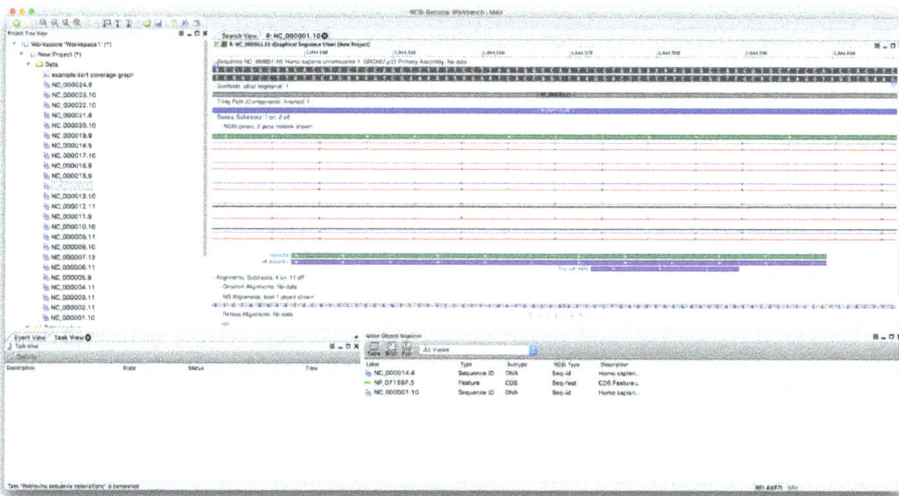

Figure 4.39 *Your results, gene annotation information and those of other projects, each one displayed as a Track, stacked up neatly. At the same time, we can interact with the display controls. Top left of the main screen is the "zoom" button which is really convenient when you need to see more detailed information or gene location to the level of the nucleotide base*

CHAPTER 5

Speeding-up with GPUs

5.1 Computational Advantages of the Graphics Card

Almost every computer has a processor component which is used to output images to a computer screen. In the 1990s, these were popularly known to experts and the layman as "graphics cards". But today, as the products of one of the most popular graphics card producer NVIDIA are known as, we have switched to referring to them as graphics processor unit (GPU).

Thus, compared to the central processor which mainly deals with the operation of the entire system, the graphics card is mainly responsible for the graphics output. Nevertheless, the computing power of the graphics card in the personal computer is increasing very rapidly relative to the processing power of the CPU, in order to serve the increasingly demanding applications regularly used by even general users. For instance, while professional users may understandably require computational power for simple image processing to high-end Hollywood animation effects, powerful graphics cards are also in demand from general users who may only play computer games or watch movies, but their appetite for speed and finer resolution and quality of the image is virtually unsatiable.

Processing an image is very complicated, of course, as the computing power of graphics cards is very low. From an actual industrial development point of view, the development of a fast graphics card has a new product's life cycle, which is much shorter than that of the central processor.

Two of the biggest market leaders, NVIDIA and AMD, virtually have to invent new architectures and next-generation flagship products every year. The direction of evolving technologies governing the development of graphics card is not much different from that of the CPU — increasing processor core clock speeds, increasing the number of cores, expansion of memory capacity and the addition of instruction sets specifically optimized for image processing needs. With tremendous developments in these areas, it is no wonder if there were not any breakthroughs. As for bioinformatics, it is no surprise that GPUs can help in improving the speed and efficacy of its many computational challenges.

From each of the technological perspectives mentioned, the focus is on the core part of the processor, including the speed and number of cores as two key priorities. On the part of clock speed, indeed there has been no significant progress and development with processor and circuit design. At the time of writing this book, NVIDIA released its Pascal architecture, the NVIDIA GeForce 10 series, with its most high-end GTX1080Ti. GTX1080Ti cores clocking in at only 1480 MHz and 1582 MHz after boost, respectively, and compared to the CPU frequently at 2 GHz to 3 GHz, it is nothing to be proud of. GTX1080Ti at its best, only hits the level of 1.5 GHz, while the 2 GHz CPU has been around for more than a dozen years.

However, if we look at the number of cores, you will feel the computing power of graphics cards is stunning. GTX1080Ti possesses 3,584 cores respectively, far far ahead of the CPU cores available. Most people can just afford to buy great six-core machines, and through HT virtualization technology turn them into 12 cores. Higher-end servers can routinely have two or four processors, and at most, up to 24 or 48 cores. In a practical sense, just like today, although we cannot afford a very clever top expert to help with our calculations, but certainly we may successfully recruit the help of more than several above-average professional. Finally, the time to get the answer would be far quicker than from a few invited experts. Hence the performance of a whole bunch of the ordinary is truly strength in numbers at work.

Concerning the standard procurement of computer hardware, it is common practice to buy super multi-core commodity servers at pretty high cost. Taking the Intel platform as an example, we need to buy more than a few commercial grade Xeon processors, each of which is currently several times more powerful than the average consumer grade Core i7 processor. Not to forget that we need to install them on as many motherboards, and this usually costs tens of thousands, not

Figure 5.1 *The main purpose of Tesla graphics cards is to compute. (Photo credit: NVIDIA.)*

something where we can dash off the nearest computer super-mall and buy one. Next, the server workstation-class power supply sets us back another five thousand US dollars. So, when we talk about the rest of the peripherals, the overall cost adds up to an amazingly huge sum.

Back to the issue of graphics card, if you buy the GTX1080Ti, this high-end consumer graphics cards merely costs US$ 999 each. The cost is thus a fraction of the money spent on high-end commercial servers. Therefore NVIDIA's TESLA product line is an appealing proposition, which enhances the overall computing power by increasing memory via the graphics card and relies on the graphics engine's higher number of cores and clock speed, each costing probably a little more than US$3,500. Compared to acquiring "supercomputers", it is not as expensive.

In fact, across the globe, there are many top supercomputing centres which depend on GPUs; by buying a huge number of graphics cards, they enhance the computing capability to petaflops scale. Thus, in the near future, the importance of graphics computing will increase. Graphics coprocessor cards would have established their case in scientific computing, and they will also feature increasingly more prominently in hardware solutions for bioinformatics.

5.2 Industry Standards and Usage of GPU Computing

Currently in the world, there are just these two brands: the NVIDIA graphics card and that of AMD (formerly ATi, later acquired by AMD). Although both companies compete intensely in the graphics card arena, they also realize that the computing power of GPUs can be used in general purpose computing outside the realm of graphics computing, and that the potential for growth is tremendous. In order to encourage developers or researchers to exploit the computing power of graphics cards more readily, they have provided a standard library of functions to facilitate general purpose GPU programming.

AMD's major development of the Radeon represents a milestone in standards, but no one denies that the whole momentum is largely driven by NVIDIA's CUDA standards. AMD did not aggressively promote this application on the market as support for its graphics coprocessor software, and seems to have forgotten about RadeonPro, resulting in a one-sided race for CUDA. (But lots of cryptocurrency miners using AMD's graphic card for specific blockchain mining, like Ethereum.)

CUDA is short for Compute Unified Device Architecture, mainly responsible for underpinning the development of specifications for NVIDIA. The emphasis is on C-language developers to use graphics architecture to enhance computing power. Thus, if researchers write in the C programming language, they can almost immediately be able to migrate existing code to run on CUDA.

As at the end of 2017, NVIDIA has developed CUDA to version 9.0 production release, with the overall pace of development considered very fast. It poses a huge challenge for timely updates and revisions by the application developer, as backward compatibility becomes a very big problem. Perhaps because of market strategy and product sales consideration, the latest version of CUDA usually only supports the latest graphics card architecture, or at best, the previous generation of products.

Maybe there is simply no choice, for the graphics card life cycle is way too short to be exhaustively backward compatible with legacy development standards. There is bound to be trade-offs in the development of new hardware architecture that is not conducive to optimization.

Even with the older program versions, appropriate modifications to make them compatible with the new version of CUDA are needed to ensure better execution efficiency and performance. With such a fast pace of development, developers must constantly face the challenge of catching up with the new standards.

To determine whether you have the correct implementation of the CUDA graphics card, and whether it functions properly, you can perform the developer test kit within an applet (these small programs are generated during the installation process of the official NVIDIA NVCC compiler), for example DeviceQuery and BandwidthTest. The former is intended to detect the NVIDIA graphics card and drivers to ensure they are correctly installed. The latter can test the data transfer bandwidth of the graphics card, and this can be a useful tool for graphics card performance assessment.

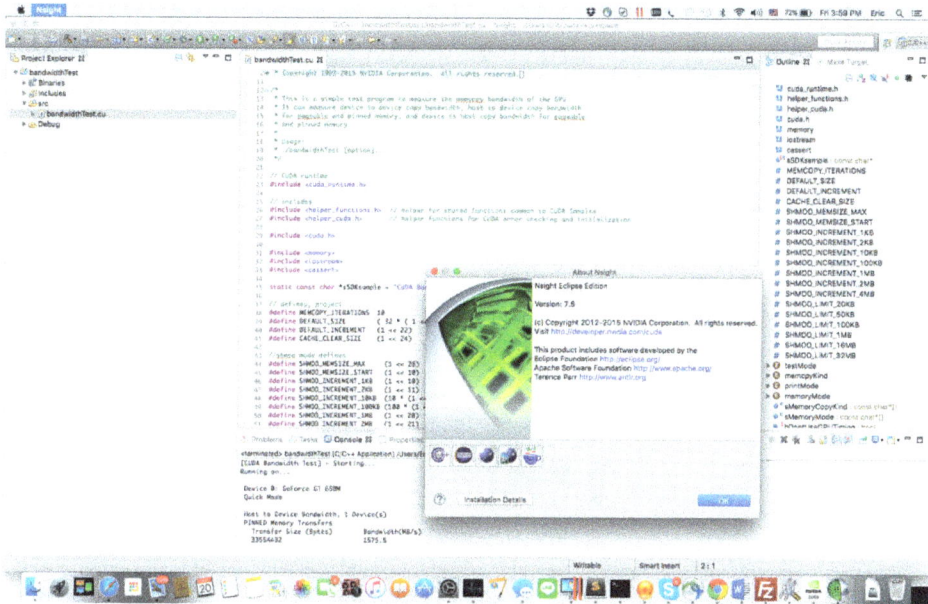

Figure 5.2 *Nsight was launched alongside NVIDIA CUDA 6.5 or 7.0 as the integrated development environment (IDE) for CUDA developers to directly edit CUDA code, and then directly perform testing and debugging. To use CUDA, the first step is to get a graphics card that supports CUDA technology, typically any NVIDIA products that is released in the past two years. Then installed on a supported operating system supports CUDA driver, the final developer kit setup is complete. In this way, you can use CUDA technology to accelerate your program*

The first is the driver installation and environment setup. For this, the version matching the operating system must be installed, particularly for the Linux platform. For Mac users, please refer to the following chart details. In the previous section, we have hinted that if you want to use CUDA, our recommendation would be to go for the Ubuntu platform. But Ubuntu is revised twice every year, usually in April and October. For example, released in October 2012, the version number of this release is 12.10. Similarly, the April 2014 version number is 14.04. Official assurances from Ubuntu that at least versions from three years will be added to LTS, for example, 14.04 LTS, for long term support.

NVIDIA official drivers are usually "limited edition." Take the 14.04 driver version installed in Ubuntu 16.10 is not necessarily usable, and for usable drivers, there is no guarantee there will not be problems during run-time. Ubuntu diehard users who habitually upgrade the system version, may actually fail to establish a good CUDA environment because of this. When this happens, we have to reinstall the operating system version corresponding to the drivers, and the software

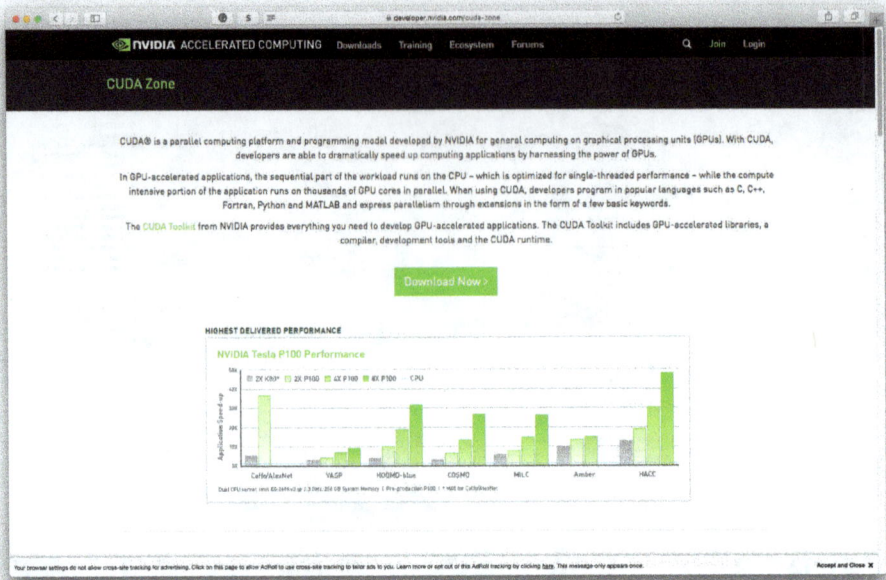

Figure 5.3 *The NVIDIA CUDA Developer Zone Web site provides a complete development kit including the CUDA driver and the program package. There are different toolkits downloadable for different operating system architectures, so be careful to download the correct one for your system*

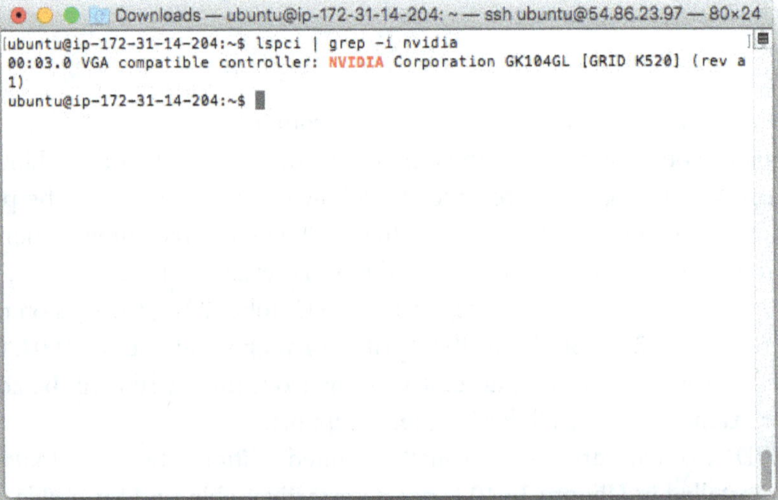

Figure 5.4 *To build the CUDA environment, for example, on the Ubuntu 14.04 platform, first check the graphic card or GPU card detected by operating system correctly, use the command:* `lspci | grep -i nvidia`

```
● ◎ ● ⌦ Downloads — ubuntu@ip-172-31-14-204: ~ — ssh ubuntu@54.86.23.97 — 80×24
ubuntu@ip-172-31-14-204:~$ sudo apt-get upgrade
Reading package lists... Done
Building dependency tree
Reading state information... Done
Calculating upgrade... Done
The following packages have been kept back:
  linux-headers-generic linux-headers-virtual linux-image-virtual
  linux-virtual
The following packages will be upgraded:
  accountsservice apparmor apport apt apt-transport-https apt-utils base-files
  bash-completion bind9-host bsdutils cloud-init curl dh-python dnsutils dpkg
  e2fslibs e2fsprogs fuse gcc-4.8-base gdisk gnupg gpgv grub-common
  grub-legacy-ec2 grub-pc grub-pc-bin grub2-common initscripts iproute2
  irqbalance isc-dhcp-client isc-dhcp-common krb5-locales libaccountsservice0
  libapparmor-perl libapparmor1 libapt-inst1.5 libapt-pkg4.12 libbind9-90
  libblkid1 libcgmanager0 libcomerr2 libcurl3 libcurl3-gnutls libdns100
  libdrm2 libexpat1 libffi6 libfreetype6 libfuse2 libgcrypt11 libgssapi-krb5-2
  libicu52 libisc95 libisccc90 libisccfg90 libk5crypto3 libkrb5-3
  libkrb5support0 libldap-2.4-2 liblwres90 libmount1 libnuma1 libpam-systemd
  libparted0debian1 libpcre3 libpng12-0 libpolkit-agent-1-0
  libpolkit-backend-1-0 libpolkit-gobject-1-0 libpython2.7
  libpython2.7-minimal libpython2.7-stdlib libpython3.4-minimal
  libpython3.4-stdlib libsqlite3-0 libss2 libssl1.0.0 libstdc++6
  libsystemd-daemon0 libsystemd-login0 libtasn1-6 libudev1 libuuid1 libxext6
```

Figure 5.5 *Before installing the official CUDA drivers and related packages, upgrade the operating system first. Please execute:* sudo apt-get update *and* sudo apt-get upgrade

```
● ◎ ● ⌦ Downloads — ubuntu@ip-172-31-14-204: ~ — ssh ubuntu@54.86.23.97 — 80×24
ubuntu@ip-172-31-14-204:~$ sudo apt-get install build-essential
Reading package lists... Done
Building dependency tree
Reading state information... Done
The following extra packages will be installed:
  binutils cpp cpp-4.8 dpkg-dev fakeroot g++ g++-4.8 gcc gcc-4.8
  libalgorithm-diff-perl libalgorithm-diff-xs-perl libalgorithm-merge-perl
  libasan0 libatomic1 libc-dev-bin libc6-dev libcloog-isl4 libdpkg-perl
  libfakeroot libfile-fcntllock-perl libgcc-4.8-dev libgmp10 libgomp1 libisl10
  libitm1 libmpc3 libmpfr4 libquadmath0 libstdc++-4.8-dev libtsan0
  linux-libc-dev make manpages-dev
Suggested packages:
  binutils-doc cpp-doc gcc-4.8-locales debian-keyring g++-multilib
  g++-4.8-multilib gcc-4.8-doc libstdc++6-4.8-dbg gcc-multilib autoconf
  automake1.9 libtool flex bison gdb gcc-doc gcc-4.8-multilib libgcc1-dbg
  libgomp1-dbg libitm1-dbg libatomic1-dbg libasan0-dbg libtsan0-dbg
  libquadmath0-dbg glibc-doc libstdc++-4.8-doc make-doc
The following NEW packages will be installed:
  binutils build-essential cpp cpp-4.8 dpkg-dev fakeroot g++ g++-4.8 gcc
  gcc-4.8 libalgorithm-diff-perl libalgorithm-diff-xs-perl
  libalgorithm-merge-perl libasan0 libatomic1 libc-dev-bin libc6-dev
  libcloog-isl4 libdpkg-perl libfakeroot libfile-fcntllock-perl libgcc-4.8-dev
  libgmp10 libgomp1 libisl10 libitm1 libmpc3 libmpfr4 libquadmath0
  libstdc++-4.8-dev libtsan0 linux-libc-dev make manpages-dev
```

Figure 5.6 *Install build-essential package (*sudo apt-get install build-essential*), the CUDA package requires the gcc compiler and related libraries*

```
● ◎ ●  ⚙ eric — ubuntu@ip-172-31-2-180: ~ — ssh ubuntu@54.173.19.115 — 80×24
ubuntu@ip-172-31-2-180:~$ wget http://developer.download.nvidia.com/compute/cuda
/7.5/Prod/local_installers/cuda-repo-ubuntu1404-7-5-local_7.5-18_amd64.deb
--2015-11-22 07:37:54--  http://developer.download.nvidia.com/compute/cuda/7.5/P
rod/local_installers/cuda-repo-ubuntu1404-7-5-local_7.5-18_amd64.deb
Resolving developer.download.nvidia.com (developer.download.nvidia.com)... 184.5
1.126.89, 184.51.126.75
Connecting to developer.download.nvidia.com (developer.download.nvidia.com)|184.
51.126.89|:80... connected.
HTTP request sent, awaiting response... 200 OK
Length: 2013382446 (1.9G) [application/x-deb]
Saving to: 'cuda-repo-ubuntu1404-7-5-local_7.5-18_amd64.deb'

100%[===================================>] 2,013,382,446  104MB/s   in 18s

2015-11-22 07:38:12 (109 MB/s) - 'cuda-repo-ubuntu1404-7-5-local_7.5-18_amd64.de
b' saved [2013382446/2013382446]

ubuntu@ip-172-31-2-180:~$ █
```

Figure 5.7 *Download the correct CUDA package for your platform, our example is CUDA 7.5 and Ubuntu 14.04, download the *.deb installer by* wget *command*

```
● ◎ ●  ⚙ eric — ubuntu@ip-172-31-2-180: ~ — ssh ubuntu@54.173.19.115 — 80×24
ubuntu@ip-172-31-2-180:~$ sudo dpkg -i cuda-repo-ubuntu1404-7-5-local_7.5-18_amd
64.deb
Selecting previously unselected package cuda-repo-ubuntu1404-7-5-local.
(Reading database ... 144882 files and directories currently installed.)
Preparing to unpack cuda-repo-ubuntu1404-7-5-local_7.5-18_amd64.deb ...
Unpacking cuda-repo-ubuntu1404-7-5-local (7.5-18) ...
Setting up cuda-repo-ubuntu1404-7-5-local (7.5-18) ...
OK
ubuntu@ip-172-31-2-180:~$ █
```

Figure 5.8 *Run the* dpkg -i <downloaded installer file> *to install the CUDA package, you may need the sudo privilege for installation*

```
●  ◎  ●  ⚷ eric — ubuntu@ip-172-31-2-180: ~ — ssh ubuntu@54.173.19.115 — 80×24
ubuntu@ip-172-31-2-180:~$ sudo apt-get install cuda
Reading package lists... Done
Building dependency tree
Reading state information... Done
The following extra packages will be installed:
  ca-certificates-java cuda-7-5 cuda-command-line-tools-7-5 cuda-core-7-5
  cuda-cublas-7-5 cuda-cublas-dev-7-5 cuda-cudart-7-5 cuda-cudart-dev-7-5
  cuda-cufft-7-5 cuda-cufft-dev-7-5 cuda-curand-7-5 cuda-curand-dev-7-5
  cuda-cusolver-7-5 cuda-cusolver-dev-7-5 cuda-cusparse-7-5
  cuda-cusparse-dev-7-5 cuda-documentation-7-5 cuda-driver-dev-7-5
  cuda-drivers cuda-license-7-5 cuda-misc-headers-7-5 cuda-npp-7-5
  cuda-npp-dev-7-5 cuda-nvrtc-7-5 cuda-nvrtc-dev-7-5 cuda-runtime-7-5
  cuda-samples-7-5 cuda-toolkit-7-5 cuda-visual-tools-7-5 default-jre
  default-jre-headless fonts-dejavu-extra freeglut3 freeglut3-dev java-common
  libatk-wrapper-java libatk-wrapper-java-jni libbonobo2-0 libbonobo2-common
  libcuda1-352 libdrm-amdgpu1 libdrm-dev libdrm-intel1 libdrm-nouveau2
  libdrm-radeon1 libdrm2 libgconf2-4 libgif4 libgl1-mesa-dev libglu1-mesa-dev
  libgnome2-0 libgnome2-bin libgnome2-common libgnomevfs2-0
  libgnomevfs2-common libice-dev libidl-common libidl0 libjansson4
  liborbit-2-0 liborbit2 libpthread-stubs0-dev libsctp1 libsm-dev libx11-dev
  libx11-doc libx11-xcb-dev libxau-dev libxcb-dri2-0-dev libxcb-dri3-dev
  libxcb-glx0-dev libxcb-present-dev libxcb-randr0-dev libxcb-render0-dev
  libxcb-shape0-dev libxcb-sync-dev libxcb-xfixes0-dev libxcb1-dev
  libxdamage-dev libxdmcp-dev libxext-dev libxfixes-dev libxi-dev libxmu-dev
```

Figure 5.9 Finish the CUDA package installation, install the CUDA related libraries and tools. Command is sudo apt-get install cuda

```
●  ◎  ●  ⚷ eric — ubuntu@ip-172-31-2-180: ~ — ssh ubuntu@54.173.19.115 — 80×24
  GNU nano 2.2.6          File: /home/ubuntu/.bashrc

# ~/.bashrc: executed by bash(1) for non-login shells.
# see /usr/share/doc/bash/examples/startup-files (in the package bash-doc)
# for examples


export CUDA_HOME=/usr/local/cuda-7.5
export LD_LIBRARY_PATH=${CUDA_HOME}/lib64

PATH=${CUDA_HOME}/bin:${PATH}
export PATH



# If not running interactively, don't do anything
case $- in
    *i*) ;;
      *) return;;
esac

                        [ Read 125 lines ]
^G Get Help  ^O WriteOut  ^R Read File  ^Y Prev Page  ^K Cut Text   ^C Cur Pos
^X Exit      ^J Justify   ^W Where Is   ^V Next Page  ^U UnCut Text ^T To Spell
```

Figure 5.10 After installation is complete, remember to create the CUDA library path environment variable. Use the UNIX export command but note that the parameters differ depending on the operating system (please refer to NVIDIA installation instruction file). In this way, the appropriate libraries can be successfully called

```
● ○ ●  ⓐ eric — ubuntu@ip-172-31-2-180: ~/NVIDIA_CUDA-7.5_Samples/1_Utilities/device...
ubuntu@ip-172-31-2-180:~$ cuda-install-samples-7.5.sh .
Copying samples to ./NVIDIA_CUDA-7.5_Samples now...
Finished copying samples.
ubuntu@ip-172-31-2-180:~$ cd NVIDIA_CUDA-7.5_Samples/1_Utilities/deviceQuery
ubuntu@ip-172-31-2-180:~/NVIDIA_CUDA-7.5_Samples/1_Utilities/deviceQuery$ make
/usr/local/cuda-7.5/bin/nvcc -ccbin g++ -I../../common/inc   -m64     -gencode arc
h=compute_20,code=sm_20 -gencode arch=compute_30,code=sm_30 -gencode arch=comput
e_35,code=sm_35 -gencode arch=compute_37,code=sm_37 -gencode arch=compute_50,cod
e=sm_50 -gencode arch=compute_52,code=sm_52 -gencode arch=compute_52,code=comput
e_52 -o deviceQuery.o -c deviceQuery.cpp
/usr/local/cuda-7.5/bin/nvcc -ccbin g++    -m64     -gencode arch=compute_20,cod
e=sm_20 -gencode arch=compute_30,code=sm_30 -gencode arch=compute_35,code=sm_35
-gencode arch=compute_37,code=sm_37 -gencode arch=compute_50,code=sm_50 -gencode
 arch=compute_52,code=sm_52 -gencode arch=compute_52,code=compute_52 -o deviceQu
ery deviceQuery.o
mkdir -p ../../bin/x86_64/linux/release
cp deviceQuery ../../bin/x86_64/linux/release
ubuntu@ip-172-31-2-180:~/NVIDIA_CUDA-7.5_Samples/1_Utilities/deviceQuery$ ▌
```

Figure 5.11   *We try to compile the sample codes from NVIDIA, copy codes to user home directory by executing the installation command:* `cuda-install-samples-7.5.sh` *(your installed version may be different). Then change current path to the sample directory, the sub-folder called 1_Utilities and its sub-folder deviceQuery. Type* `make` *and hit* `Enter` *to compile the deviceQuery, a CUDA application example*

```
● ○ ●  ⓐ eric — ubuntu@ip-172-31-2-180: ~/NVIDIA_CUDA-7.5_Samples/1_Utilities/device...
ubuntu@ip-172-31-2-180:~/NVIDIA_CUDA-7.5_Samples/1_Utilities/deviceQuery$ ./devi
ceQuery
./deviceQuery Starting...

 CUDA Device Query (Runtime API) version (CUDART static linking)

Detected 1 CUDA Capable device(s)

Device 0: "GRID K520"
  CUDA Driver Version / Runtime Version          7.5 / 7.5
  CUDA Capability Major/Minor version number:    3.0
  Total amount of global memory:                 4096 MBytes (4294770688 bytes)
  ( 8) Multiprocessors, (192) CUDA Cores/MP:     1536 CUDA Cores
  GPU Max Clock rate:                            797 MHz (0.80 GHz)
  Memory Clock rate:                             2500 Mhz
  Memory Bus Width:                              256-bit
  L2 Cache Size:                                 524288 bytes
  Maximum Texture Dimension Size (x,y,z)         1D=(65536), 2D=(65536, 65536),
3D=(4096, 4096, 4096)
  Maximum Layered 1D Texture Size, (num) layers  1D=(16384), 2048 layers
  Maximum Layered 2D Texture Size, (num) layers  2D=(16384, 16384), 2048 layers
  Total amount of constant memory:               65536 bytes
  Total amount of shared memory per block:       49152 bytes
  Total number of registers available per block: 65536
```

Figure 5.12   *Finish the compilation. Running* `./deviceQuery` *will see current CUDA device and library version information, this step can confirm the environment setup is correct*

```
● ◌ ● ⟨ eric — ubuntu@ip-172-31-2-180: ~/NVIDIA_CUDA-7.5_Samples/1_Utilities/bandwi...
ubuntu@ip-172-31-2-180:~/NVIDIA_CUDA-7.5_Samples/1_Utilities/bandwidthTest$ ls
Makefile  NsightEclipse.xml  bandwidthTest.cu  readme.txt
ubuntu@ip-172-31-2-180:~/NVIDIA_CUDA-7.5_Samples/1_Utilities/bandwidthTest$ make
/usr/local/cuda-7.5/bin/nvcc -ccbin g++ -I../../common/inc  -m64     -gencode arc
h=compute_20,code=sm_20 -gencode arch=compute_30,code=sm_30 -gencode arch=comput
e_35,code=sm_35 -gencode arch=compute_37,code=sm_37 -gencode arch=compute_50,cod
e=sm_50 -gencode arch=compute_52,code=sm_52 -gencode arch=compute_52,code=comput
e_52 -o bandwidthTest.o -c bandwidthTest.cu
/usr/local/cuda-7.5/bin/nvcc -ccbin g++   -m64      -gencode arch=compute_20,cod
e=sm_20 -gencode arch=compute_30,code=sm_30 -gencode arch=compute_35,code=sm_35
-gencode arch=compute_37,code=sm_37 -gencode arch=compute_50,code=sm_50 -gencode
 arch=compute_52,code=sm_52 -gencode arch=compute_52,code=compute_52 -o bandwidt
hTest bandwidthTest.o
mkdir -p ../../bin/x86_64/linux/release
cp bandwidthTest ../../bin/x86_64/linux/release
ubuntu@ip-172-31-2-180:~/NVIDIA_CUDA-7.5_Samples/1_Utilities/bandwidthTest$ ▊
```

Figure 5.13 *Except deviceQuery, you can find bandwidthTest in upper folder, please use* make *command to compile it*

```
● ◌ ● ⟨ eric — ubuntu@ip-172-31-2-180: ~/NVIDIA_CUDA-7.5_Samples/1_Utilities/bandwi...
ubuntu@ip-172-31-2-180:~/NVIDIA_CUDA-7.5_Samples/1_Utilities/bandwidthTest$ ./ba
ndwidthTest
[CUDA Bandwidth Test] - Starting...
Running on...

 Device 0: GRID K520
 Quick Mode

 Host to Device Bandwidth, 1 Device(s)
 PINNED Memory Transfers
   Transfer Size (Bytes)        Bandwidth(MB/s)
   33554432                     9115.1

 Device to Host Bandwidth, 1 Device(s)
 PINNED Memory Transfers
   Transfer Size (Bytes)        Bandwidth(MB/s)
   33554432                     8189.5

 Device to Device Bandwidth, 1 Device(s)
 PINNED Memory Transfers
   Transfer Size (Bytes)        Bandwidth(MB/s)
   33554432                     118907.3

Result = PASS
```

Figure 5.14 *This tool, bandwidthTest, tests the graphics card bandwidth. From this, we can clearly understand the CUDA graphics card memory, the data transmission capabilities and graphics card internal data transfer rates. The above description of the graphics card gives us an idea of the process involved, which looks simple, but in practice, there are many problems to overcome*

```
● ● ●   ubuntu@ip-172-31-51-147: ~/NVIDIA_CUDA-7.5_Samples — ssh ubuntu@52.91.78.2...
[ubuntu@ip-172-31-51-147:~/NVIDIA_CUDA-7.5_Samples$ make
make[1]: Entering directory `/home/ubuntu/NVIDIA_CUDA-7.5_Samples/0_Simple/cdpSi
mplePrint'
/usr/local/cuda-7.5/bin/nvcc -ccbin g++ -I../../common/inc  -m64    -dc -gencode
 arch=compute_35,code=sm_35 -gencode arch=compute_37,code=sm_37 -gencode arch=co
mpute_50,code=sm_50 -gencode arch=compute_52,code=sm_52 -gencode arch=compute_52
,code=compute_52 -o cdpSimplePrint.o -c cdpSimplePrint.cu
/usr/local/cuda-7.5/bin/nvcc -ccbin g++    -m64       -gencode arch=compute_35,cod
e=sm_35 -gencode arch=compute_37,code=sm_37 -gencode arch=compute_50,code=sm_50
-gencode arch=compute_52,code=sm_52 -gencode arch=compute_52,code=compute_52 -o
cdpSimplePrint cdpSimplePrint.o  -lcudadevrt
mkdir -p ../../bin/x86_64/linux/release
cp cdpSimplePrint ../../bin/x86_64/linux/release
make[1]: Leaving directory `/home/ubuntu/NVIDIA_CUDA-7.5_Samples/0_Simple/cdpSim
plePrint'
make[1]: Entering directory `/home/ubuntu/NVIDIA_CUDA-7.5_Samples/0_Simple/matri
xMulDrv'
/usr/local/cuda-7.5/bin/nvcc -ccbin g++ -I../../common/inc  -m64     -gencode arc
h=compute_20,code=compute_20 -o matrixMulDrv.o -c matrixMulDrv.cpp
/usr/local/cuda-7.5/bin/nvcc -ccbin g++    -m64        -gencode arch=compute_20,cod
e=compute_20 -o matrixMulDrv matrixMulDrv.o  -lcuda
mkdir -p ../../bin/x86_64/linux/release
cp matrixMulDrv ../../bin/x86_64/linux/release
/usr/local/cuda-7.5/bin/nvcc -ccbin g++ -I../../common/inc  -m64     -gencode arc
```

Figure 5.15   *Change directory to NVIDIA_CUDA-7.5_Samples, and within the folder, type "make" to start compiling all examples*

```
● ● ●   ⌂ eric — ubuntu@ip-172-31-51-147: ~/NVIDIA_CUDA-7.5_Samples/bin/x86_64/linux...
[ubuntu@ip-172-31-51-147:~/NVIDIA_CUDA-7.5_Samples/bin/x86_64/linux/release$ ls
BlackScholes              matrixMulDrv
BlackScholes_nvrtc        matrixMulDynlinkJIT
FDTD3d                    matrixMul_kernel64.ptx
FunctionPointers          matrixMul_nvrtc
HSOpticalFlow             mergeSort
MC_EstimatePiInlineP      nbody
MC_EstimatePiInlineQ      newdelete
MC_EstimatePiP            oceanFFT
MC_EstimatePiQ            p2pBandwidthLatencyTest
MC_SingleAsianOptionP     particles
Mandelbrot                postProcessGL
MersenneTwisterGP11213    ptxjit
MonteCarloMultiGPU        quasirandomGenerator
NV12ToARGB_drvapi64.ptx   quasirandomGenerator_nvrtc
SobelFilter               radixSortThrust
SobolQRNG                 randomFog
StreamPriorities          recursiveGaussian
UnifiedMemoryStreams      reduction
alignedTypes              scalarProd
asyncAPI                  scan
bandwidthTest             segmentationTreeThrust
batchCUBLAS               shfl_scan
bicubicTexture            simpleAssert
```

Figure 5.16   *Compiled programs will be placed in the NVIDIA_CUDA-7.5_Samples folder under the subfolder /bin/x86_64/linux/release. The directory listing confirms the content of Samples, the functions of which are in accordance with the program's name*

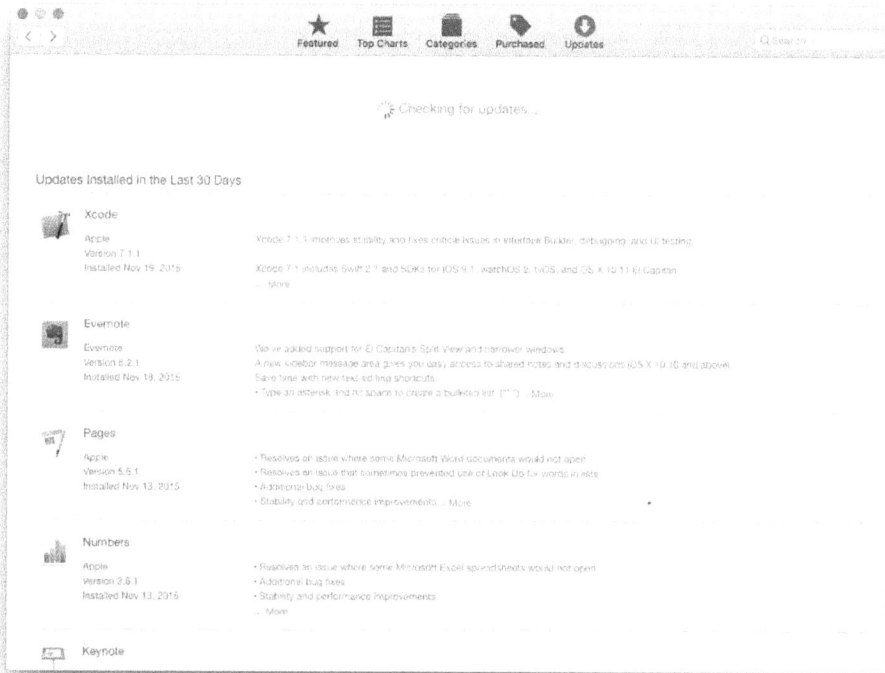

Figure 5.17 CUDA environment built on a Mac seems like the easiest, but there are some details you would still have to pay attention. For 10.11 El Captain, for example, we can install from App Store, Xcode. The above example shows version 7.1

development kit (SDK) rebuilt again with the official NVCC compilation applet. My experience with a build environment like CUDA, to follow the typical policies of enterprises insofar as upgrading operating systems is concerned. Compatibility issues are a very big challenge. Hence there is no real need to pursue the latest version. Just track the reaction from the community before deciding when and whether to upgrade.

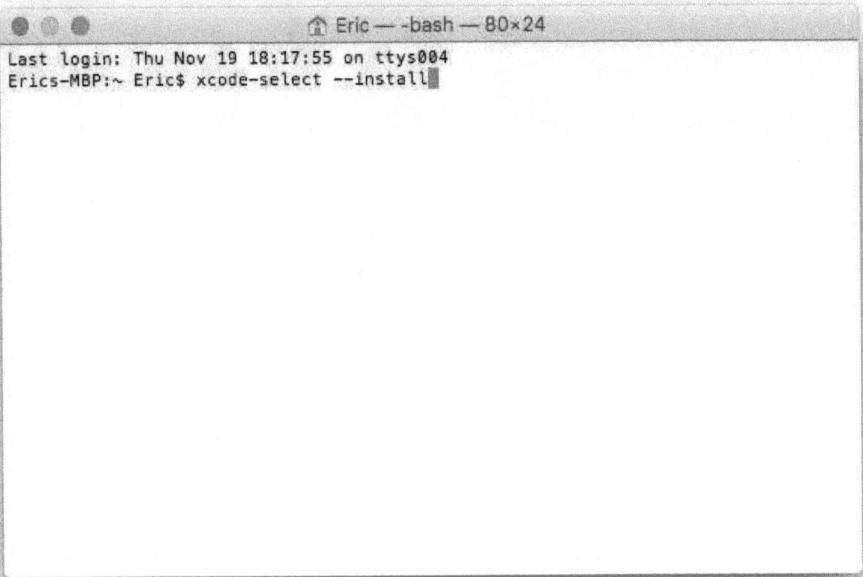

Figure 5.18 *After OS X 10.9, the installation of command line tools require user run the* "xcode-select --install" *command within the terminal*

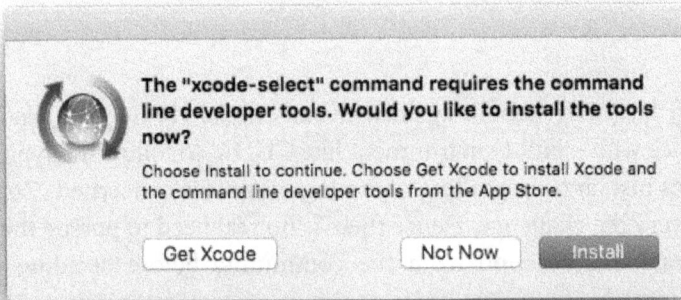

Figure 5.19 *Prompt the system dialogue window, click the* Install *button for installation*

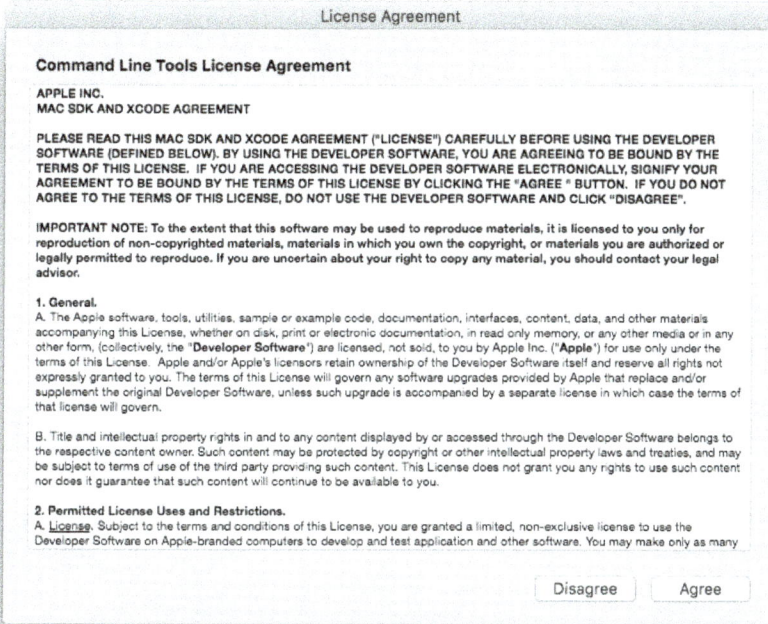

Figure 5.20 *You must accept the software agreement then continue the command line tools installation*

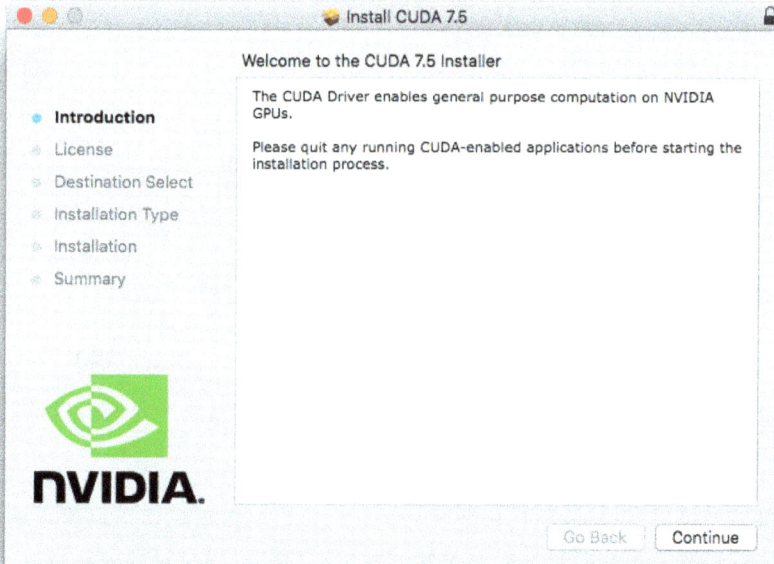

Figure 5.21 *Installing Mac version of the CUDA driver, version 7.5, suit for the macOS after build 10.9*

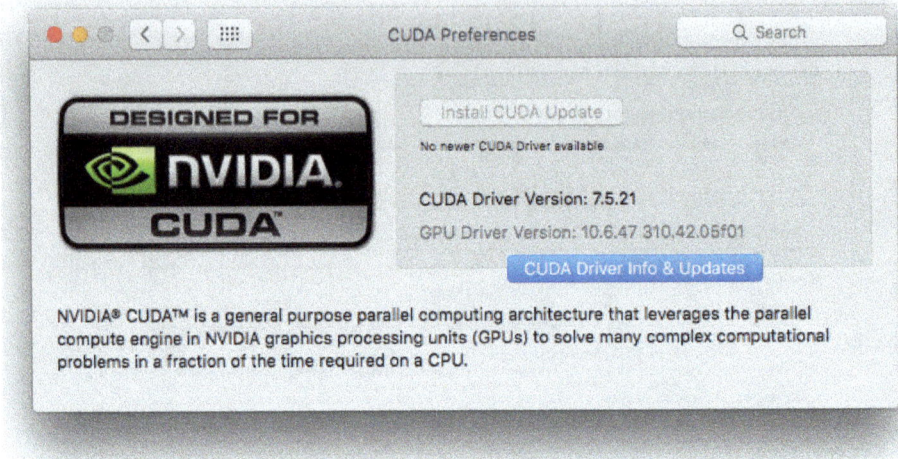

Figure 5.22 *Once the CUDA driver installation is complete, the System Preferences will contain one more option, CUDA*

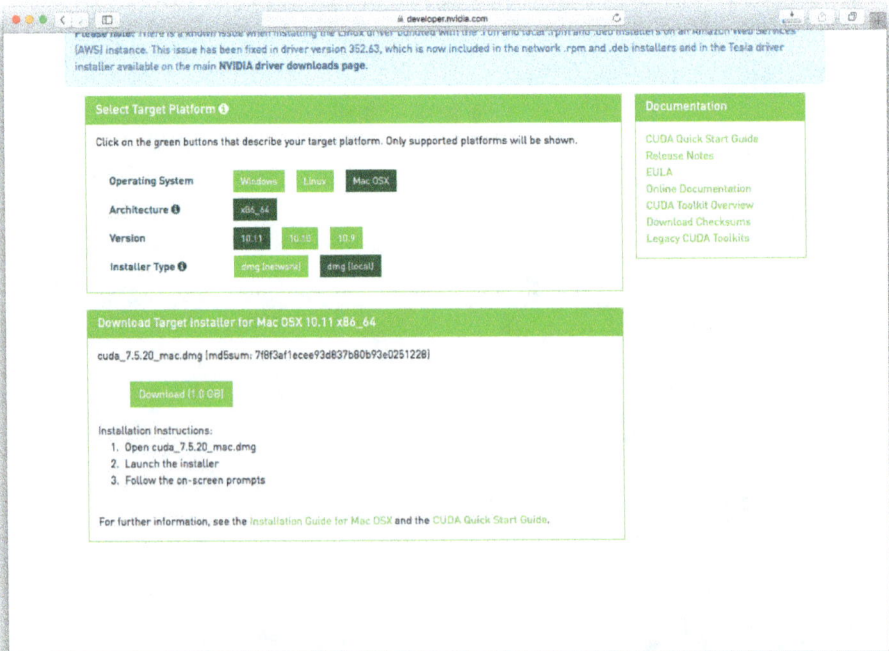

Figure 5.23 *Download the Cuda Toolkit from Nvidia website, make sure the platform and version are correct. For example, my macOS is 10.11, choose the complete installer DMG file (dmg local)*

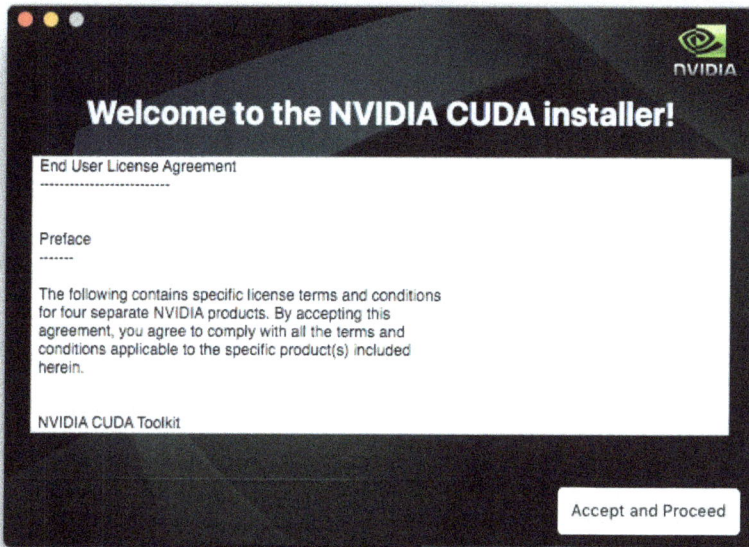

Figure 5.24 *Open the DMG file and click the installer, start to install CUDA toolkit for macOS, click* `Accept and Proceed` *button*

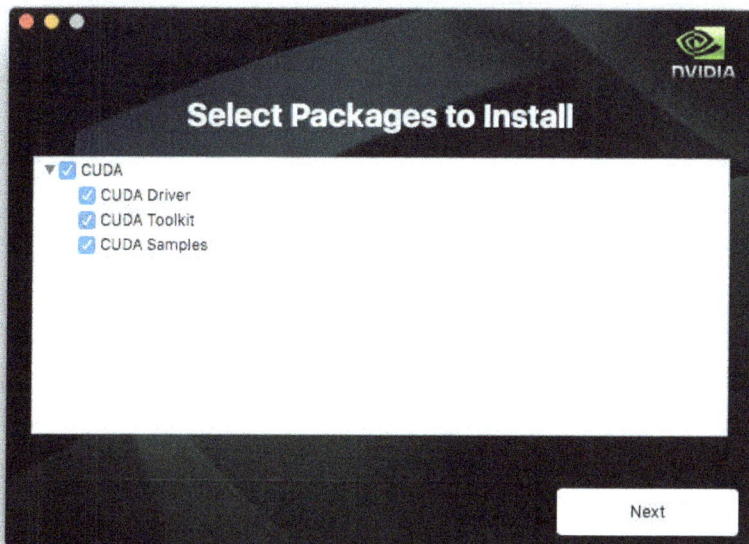

Figure 5.25 *Select all of the three packages, click "*`Next`*" to continue*

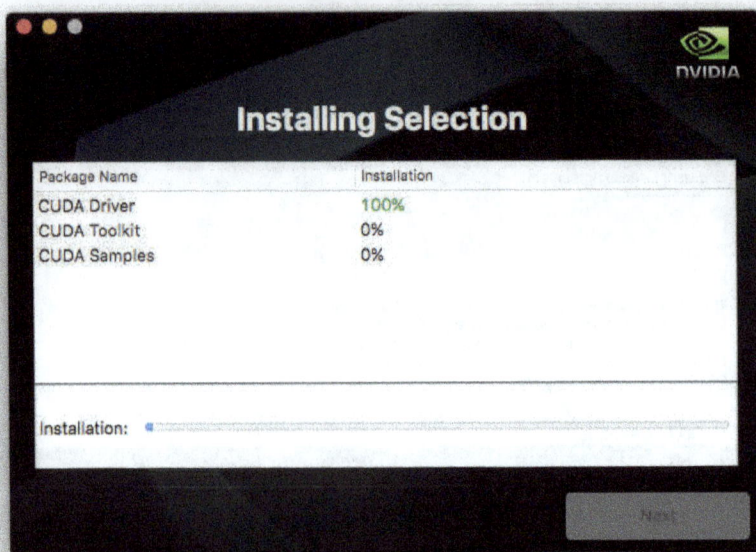

Figure 5.26 *Installation progress for three components, please wait*

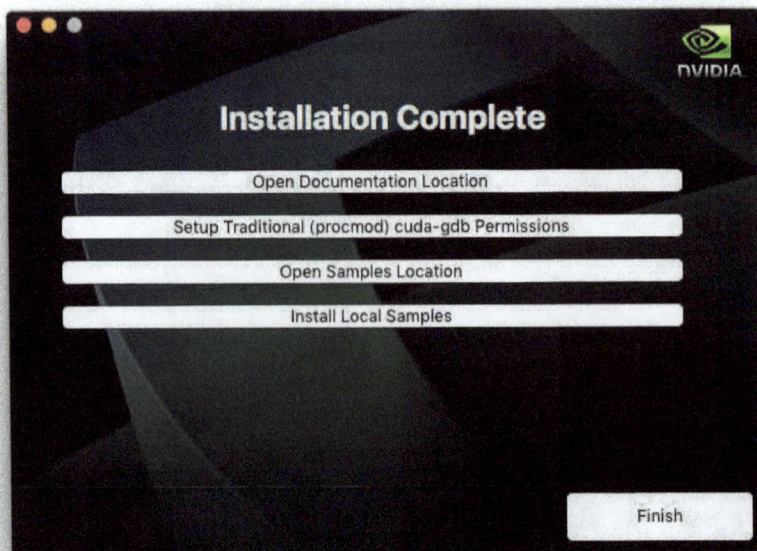

Figure 5.27 *Installation complete. Click the second item* "Setup Traditional (procmod) cuda-gdb Permissions" *and the fourth* "Install Local Samples". *The two operations help you setup permissions and copy the CUDA sample codes to your Mac*

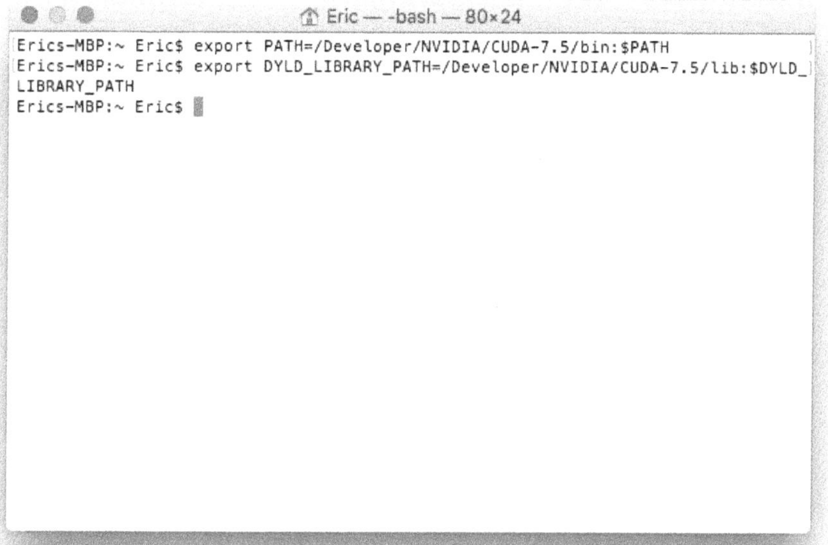

Figure 5.28 *Like the Linux version, we need to set the environment variable in Mac, execute two export commands in sequence:* `export   PATH=/Developer/NVIDIA/CUDA-7.5/bin:$PATH export DYLD_LIBRARY_PATH=/Developer/NVIDIA/CUDA-7.5/lib:$DYLD_LIBRARY_PATH`

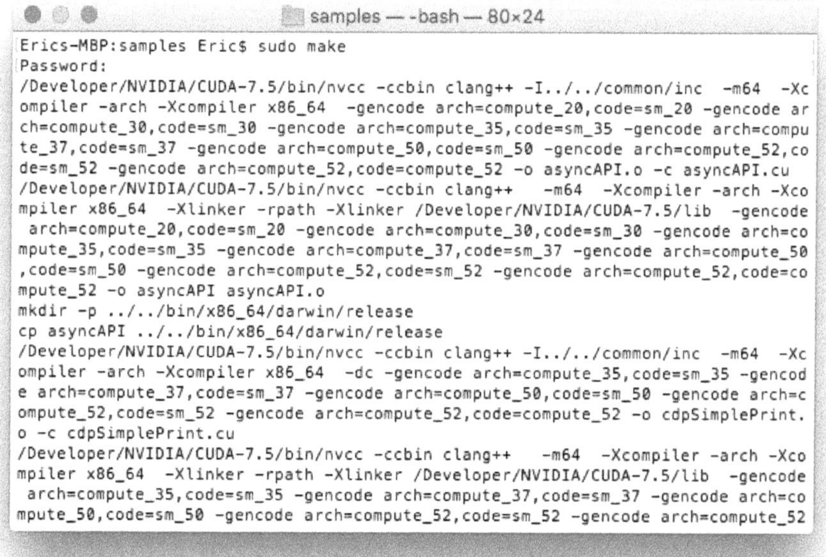

Figure 5.29 *In the directory, /Developer/NVIDIA/CUDA-7.5/samples, enter* `sudo   make`, *to start compiling the official NVIDIA CUDA Samples*

```
●  ●  ●                    📁 release — -bash — 80×24
[Erics-MBP:release Eric$ ./deviceQuery                                     ]
./deviceQuery Starting...

 CUDA Device Query (Runtime API) version (CUDART static linking)

Detected 1 CUDA Capable device(s)

Device 0: "GeForce GT 650M"
  CUDA Driver Version / Runtime Version          7.5 / 7.5
  CUDA Capability Major/Minor version number:    3.0
  Total amount of global memory:                 512 MBytes (536543232 bytes)
  ( 2) Multiprocessors, (192) CUDA Cores/MP:     384 CUDA Cores
  GPU Max Clock rate:                            405 MHz (0.41 GHz)
  Memory Clock rate:                             2000 Mhz
  Memory Bus Width:                              128-bit
  L2 Cache Size:                                 262144 bytes
  Maximum Texture Dimension Size (x,y,z)         1D=(65536), 2D=(65536, 65536),
3D=(4096, 4096, 4096)
  Maximum Layered 1D Texture Size, (num) layers  1D=(16384), 2048 layers
  Maximum Layered 2D Texture Size, (num) layers  2D=(16384, 16384), 2048 layers
  Total amount of constant memory:               65536 bytes
  Total amount of shared memory per block:       49152 bytes
  Total number of registers available per block: 65536
  Warp size:                                     32
```

Figure 5.30   *Mac version of CUDA example, will be placed after compilation in the directory,* */Developer/CUDA-7.5/samples/bin/x86_64/darwin/release. Similar to the earlier example on Linux, take a look at the installed programs for example, try* `deviceQuery`

```
●  ●  ●                    📁 release — -bash — 80×24
[Erics-MBP:release Eric$ ./bandwidthTest                                   ]
[CUDA Bandwidth Test] - Starting...
Running on...

 Device 0: GeForce GT 650M
 Quick Mode

 Host to Device Bandwidth, 1 Device(s)
 PINNED Memory Transfers
   Transfer Size (Bytes)        Bandwidth(MB/s)
   33554432                     1604.2

 Device to Host Bandwidth, 1 Device(s)
 PINNED Memory Transfers
   Transfer Size (Bytes)        Bandwidth(MB/s)
   33554432                     1778.3

 Device to Device Bandwidth, 1 Device(s)
 PINNED Memory Transfers
   Transfer Size (Bytes)        Bandwidth(MB/s)
   33554432                     20061.3

Result = PASS
```

Figure 5.31   *Then carry out the test,* `bandwidthTest`, *to make sure everything is in order, just as in the previous example with Linux*

Notebooks, because of the need for a power-saving architecture, have to be managed differently. So the operation of the graphics card is not the same as the power-saving features affect the use of CUDA.

Currently, commercially available laptops mostly use the NVIDIA Optimus architecture, because of battery power energy-saving situation in the laptop, on-board integrated graphics chips such as Intel HD Graphic, or a separate NVIDIA graphics card is used. At any one time, we can use only one chip, and there is no way to use integrated graphics, but we can access standalone discrete graphics card for CUDA. CUDA programs running on the notebook, for the moment must be using NVIDIA graphics cards. If you have already installed the CUDA driver, set up the build environment, and you execute deviceQuery, but it does not show up the CUDA device, the reason is probably this power-saving feature.

In Mac, we will have to force the use of NVIDIA graphics cards. In the System Preferences under energy-saving schemes, turn off the automatic switch-off graphics card function. This will force to use NVIDIA graphics cards. As for the Linux OS, it is not so simple. No wonder the father of Linux, Linus Torvalds, in a public lecture open dissatisfaction was expressed at NVIDIA, for failing to open the source code on Optimus switching.

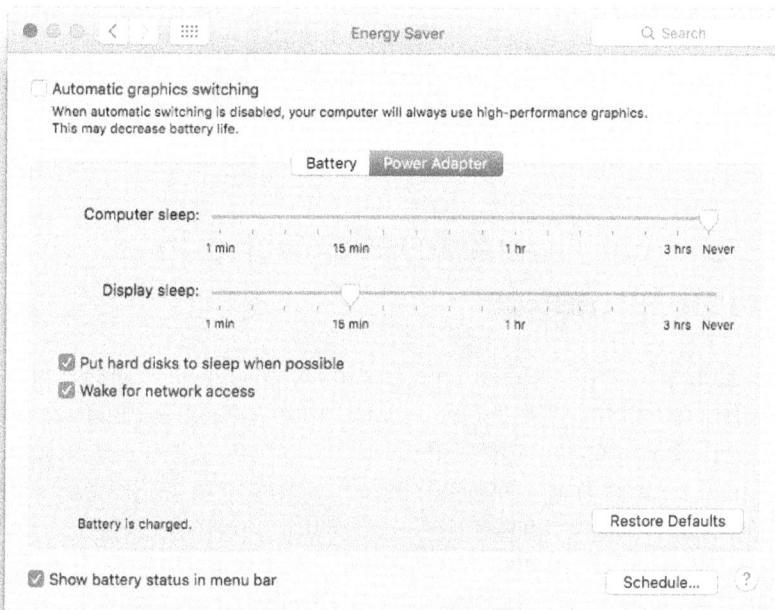

Figure 5.32 *In the Mac OS X, under System Preferences, unchecking the feature to set to automatically switch off graphics card (*Automatic graphics switching*), we can force NVIDIA graphics cards to be used*

Figure 5.33   *From macOS System Information (About This Mac) display (Displays) page, you can now see the graphics card being used*

5.3 Practical CUDA Applications in Bioinformatics

The use of graphics card has great computational advantage. The ability to access multiple cores allows for amazing parallel computing on an ordinary machine. It reduces computing time, providing rapid results. For computational biology applications, which are usually time consuming, we can theoretically provide the much-needed acceleration. In fact, there are many research projects involving bio-computing where the application of graphics cards is crucial.

Looking at the current academic papers published, these graphics applications are divided into two camps: sequence alignment and protein docking. The former has been introduced in a separate chapter on sequence alignment in bioinformatics. For protein docking, the task is to solve the problem of how small molecules bind in the correct position on the protein surface. After binding takes place, what does the structure of the complex look like, what functions are affected, and how. Researchers need to manipulate the structure, flip and rotate, and so on. Such operations for graphics simulation are ideal for computation using graphics cards.

Currently in more practical aspects of research, there are many open source tools emerging, especially on sequence alignment. Sequence alignment using the traditional Smith-Waterman algorithm typically requires huge computing resources, particularly where the datasets are large, particularly where the results require calculations using dynamic programming, to achieve precise sequence comparison.

Because of relatively large computing resources required, this is where CUDA comes in. The tools for the application of the Smith-Waterman algorithm for sequence alignment, CUSHAW2, has been developed to version 2.4. Burrows-Wheeler Transformation (BWT) suitable for large-scale sequence alignment, necessitates building an index. CUDA has CUSHAW comparison tool, which can be used. This is similar in function to the BWT algorithm, but it is a unique algorithm, which can facilitate fast and accurate SOAP. The version SOAP3 and SOAP3-dp was also introduced with CUDA computing.

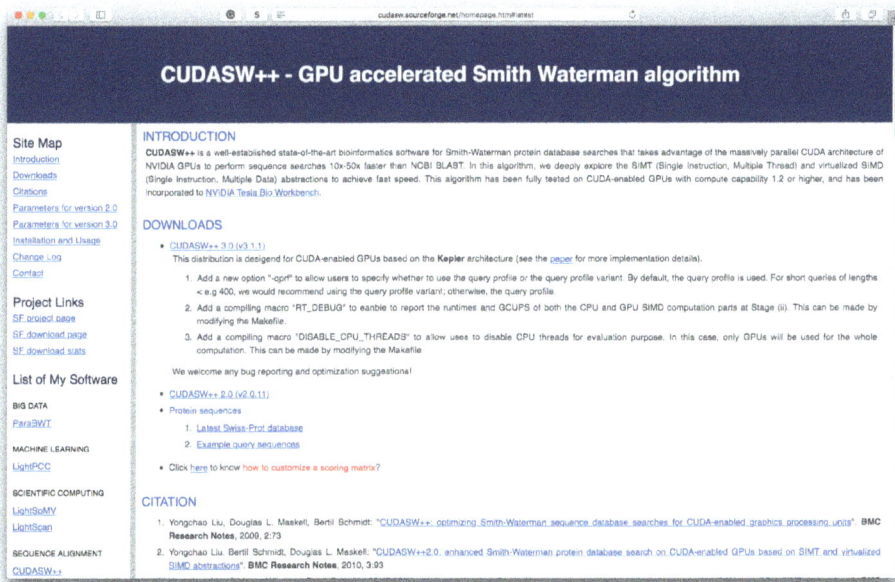

Figure 5.34 *CUDASW++ (http://cudasw.sourceforge.net) is a famous CUDA-based alignment tool. NVIDIA also integrated it to their nvBio toolkits. Not only CUDASW++, the author also provide many CUDA-based bioinformatics related tools*

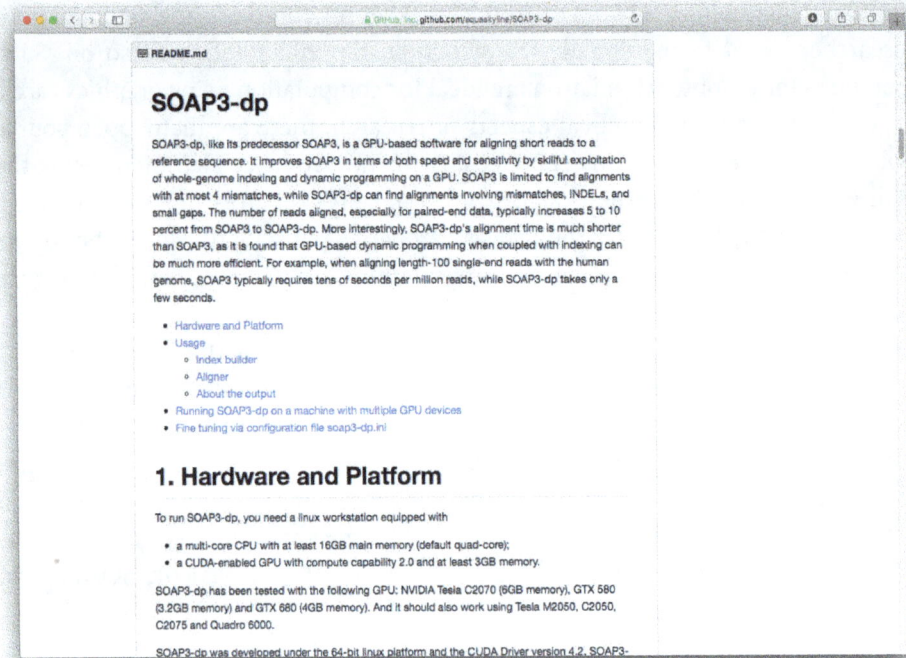

Figure 5.35   *SOAP3-dp developed by BGI, main features and algorithm inherited from SOAP3, with better performance and accuracy*

Now we want to use the created CUDA platform, actual use the CUSHAW2-GPU for sequence alignment. As the understanding of BWT, the operation including creating reference sequence file and input the FASTQ file for alignment.

Preparing the reference sequence

You can download the pre-built sequence indexing file from CUSHAW2 website. Or using CUSHAW2_INDEX to create the indexing file for HG19 FASTA file.

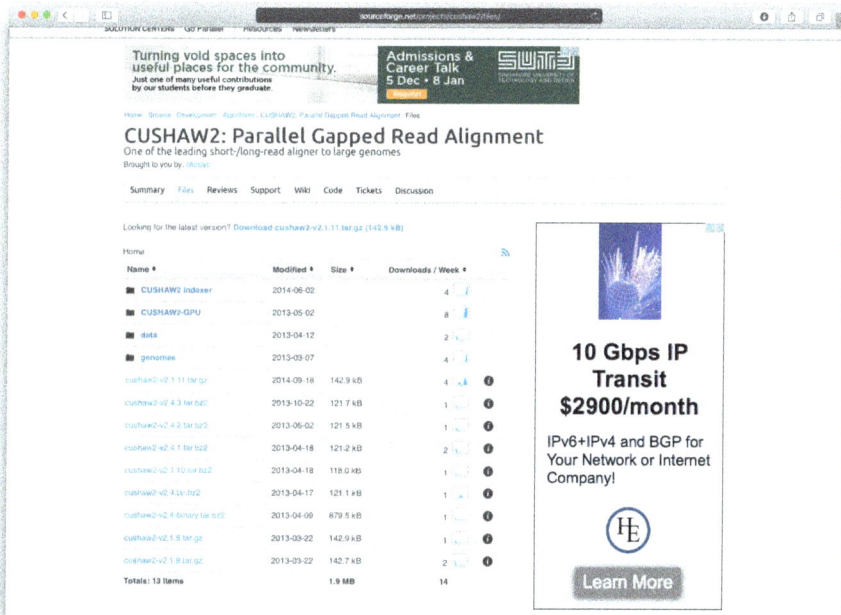

Figure 5.36 *SourceForge page for CUSHAW2, you can download the necessary tools for the demonstration later or download the pre-built indexing reference sequence files*

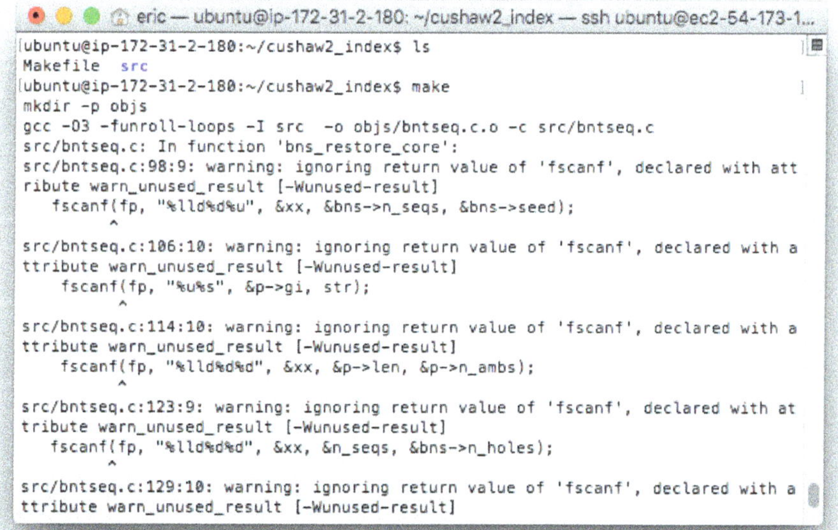

Figure 5.37 *Download the cushaw2_index source code, use* make *command for compilation*

```
● ◉ ● ⌂ eric — ubuntu@ip-172-31-2-180: /mnt — ssh ubuntu@54.173.19.115 — 80×24
ubuntu@ip-172-31-2-180:/mnt$ cushaw2_index -a bwtsw hg19.fa           ]▤
Only the reverse BWT
SA interval: 8 Orient: 2
[cushaw2_index] Pack FASTA... 30.68 sec
[cushaw2_index] Reverse the packed sequence... 9.26 sec
[cushaw2_index] Construct BWT for the reverse packed sequence...
[BWTIncConstructFromPacked] 10 iterations done. 100000000 characters processed.
```

Figure 5.38   *Finish the compilation, change the directory containing FASTA file (e.g. hg19.fa). Run the command to build reference sequence indexing file:* `cushaw2_index -a bwtsw hg19.fa`

```
● ◉ ● ⌂ eric — ubuntu@ip-172-31-2-180: /mnt — ssh ubuntu@54.173.19.115 — 80×24
[BWTIncConstructFromPacked] 230 iterations done. 2300000000 characters processed

[BWTIncConstructFromPacked] 240 iterations done. 2400000000 characters processed

[BWTIncConstructFromPacked] 250 iterations done. 2500000000 characters processed

[BWTIncConstructFromPacked] 260 iterations done. 2600000000 characters processed

[BWTIncConstructFromPacked] 270 iterations done. 2700000000 characters processed

[BWTIncConstructFromPacked] 280 iterations done. 2800000000 characters processed

[BWTIncConstructFromPacked] 290 iterations done. 2900000000 characters processed

[BWTIncConstructFromPacked] 300 iterations done. 3000000000 characters processed

[BWTIncConstructFromPacked] 310 iterations done. 3098480720 characters processed

[bwt_gen] Finished constructing BWT in 315 iterations.
[cushaw2_index] 1214.43 seconds elapse.
[cushaw2_index] Update reverse BWT... 25.93 sec
[cushaw2_index] Construct SA from reverse BWT and Occ... 421.12 sec
ubuntu@ip-172-31-2-180:/mnt$ ▮
```

Figure 5.39   *Wait for a while, need some time to build index. Our machine with 8 cores consumed 30 minutes*

Alignment with CUSHAW2-GPU

```
● ● ● 📁 ref — ubuntu@ip-172-31-51-147: /mnt/cushaw2-gpu-2.1.8-r16 — ssh ubuntu@5...
ubuntu@ip-172-31-51-147:/mnt/cushaw2-gpu-2.1.8-r16$ make
mkdir -p objs
make -C bamreader
make[1]: Entering directory `/mnt/cushaw2-gpu-2.1.8-r16/bamreader'
mkdir -p objs
gcc -static -O3 -Wall -funroll-loops -o objs/bam_aux.c.o -c bam_aux.c
gcc -static -O3 -Wall -funroll-loops -o objs/bam.c.o -c bam.c
gcc -static -O3 -Wall -funroll-loops -o objs/bam_import.c.o -c bam_import.c
gcc -static -O3 -Wall -funroll-loops -o objs/bgzf.c.o -c bgzf.c
bgzf.c: In function 'bgzf_close':
bgzf.c:619:8: warning: variable 'count' set but not used [-Wunused-but-set-varia
ble]
    int count, block_length = deflate_block(fp, 0);
        ^
bgzf.c: In function 'bgzf_check_EOF':
bgzf.c:672:7: warning: ignoring return value of 'fread', declared with attribute
 warn_unused_result [-Wunused-result]
   fread(buf, 1, 28, fp->file);
        ^
gcc -static -O3 -Wall -funroll-loops -o objs/faidx.c.o -c faidx.c
gcc -static -O3 -Wall -funroll-loops -o objs/kstring.c.o -c kstring.c
gcc -static -O3 -Wall -funroll-loops -o objs/razf.c.o -c razf.c
razf.c: In function 'razf_flush':
razf.c:243:8: warning: ignoring return value of 'write', declared with attribute
```

Figure 5.40   *Download cushaw2-gpu-2.1.8-r16 and unzip it. Use* make *for compilation*

```
● ● ● 🏠 eric — ubuntu@ip-172-31-51-147: /mnt/ref/cushaw — ssh ubuntu@52.91.78.228...
 -r ../hg19.fa -f ../../data/SRR101437.fastq -o gpu.sam
Command: ../../cushaw2-gpu-2.1.8-r16/cushaw2-gpu -r ../hg19.fa -f ../../data/SRR
101437.fastq -o gpu.sam
[P6121]#CUDA-enabled devices: 1
[P6121]---------Qualified GPU (index: 0)-------------
[P6121]name:GRID K520
[P6121]multiprocessor count:8
[P6121]clock rate:797000
[P6121]shared memory:49152
[P6121]global memory:4294770688
[P6121]registers per block:65536
[P6121]compute capability:3.0
[P6121]-----------------------------------------------
[P6121]Require GPUs with compute capability >= 3.0
[P6121]Number of Qualified GPUs: 1
[P6121]Number of extra CPU threads: 0
[P6121]GPU index: 0
[P6121]Suffix array memory size: 1495.92 MB
[P6121]load BWT (1121 MB)
[P6121]load packed genome (1495 MB)
[P6121]FASTQ format identified
[P6121]Max number of reads per batch: 65536
[P6121]processed 655360 reads by the GPU in 16.45 seconds
```

Figure 5.41   *After compilation, start to execute sequence alignment. Use* cushaw2-gpu -r *parameter for reference sequence,* -f *for FASTQ file input. Alignment output use* -o *parameter, format is SAM*

```
● ● ●   ubuntu@ip-172-31-51-147: /mnt/ref — ssh ubuntu@52.91.78.228 — 80×24
ubuntu@ip-172-31-51-147:/mnt/ref$ nvidia-smi
Sun Nov 22 12:00:29 2015
+------------------------------------------------------+
| NVIDIA-SMI 352.63      Driver Version: 352.63         |
|-------------------------------+----------------------+----------------------+
| GPU  Name        Persistence-M| Bus-Id        Disp.A | Volatile Uncorr. ECC |
| Fan  Temp  Perf  Pwr:Usage/Cap|         Memory-Usage | GPU-Util  Compute M. |
|===============================+======================+======================|
|   0  GRID K520           Off  | 0000:00:03.0     Off |                  N/A |
| N/A   42C    P0    79W / 125W |   3756MiB / 4095MiB  |     88%      Default |
+-------------------------------+----------------------+----------------------+

+-----------------------------------------------------------------------------+
| Processes:                                                       GPU Memory |
|  GPU       PID  Type  Process name                               Usage      |
|=============================================================================|
|    0      6121    C   ../../cushaw2-gpu-2.1.8-r16/cushaw2-gpu        3737MiB |
+-----------------------------------------------------------------------------+
ubuntu@ip-172-31-51-147:/mnt/ref$
```

Figure 5.42 *During the alignment process, you can query the NVIDIA graphic/CUDA card status by nvidia-smi command. See the "Processes" section, you can see the CUSHAW2-GPU execution, consumed 3.7GB GPU memory*

```
● ● ●   ⚤ eric — ubuntu@ip-172-31-51-147: /mnt/ref/cushaw — ssh ubuntu@52.91.78.228...
[P31113]processed 5242880 reads by the GPU in 123.75 seconds
[P31113]processed 5898240 reads by the GPU in 139.40 seconds
[P31113]processed 6553600 reads by the GPU in 154.68 seconds
[P31113]processed 7208960 reads by the GPU in 169.95 seconds
[P31113]processed 7864320 reads by the GPU in 185.10 seconds
[P31113]processed 8519680 reads by the GPU in 200.30 seconds
[P31113]processed 9175040 reads by the GPU in 215.65 seconds
[P31113]processed 9830400 reads by the GPU in 231.36 seconds
[P31113]processed 10485760 reads by the GPU in 246.78 seconds
[P31113]processed 11141120 reads by the GPU in 263.12 seconds
[P31113]processed 11796480 reads by the GPU in 278.78 seconds
[P31113]processed 12451840 reads by the GPU in 294.01 seconds
[P31113]processed 13107200 reads by the GPU in 309.28 seconds
[P31113]processed 13762560 reads by the GPU in 326.03 seconds
[P31113]processed 14417920 reads by the GPU in 342.37 seconds
[P31113]processed 15073280 reads by the GPU in 358.64 seconds
[P31113]processed 15728640 reads by the GPU in 374.54 seconds
[P31113]processed 16384000 reads by the GPU in 389.74 seconds
[P31113]processed 17039360 reads by the GPU in 406.89 seconds
[P31113]processed 17694720 reads by the GPU in 422.55 seconds
[P31113]processed 18350080 reads by the GPU in 438.03 seconds
[P31113]#Reads aligned: 18134294 / 18466358 (98.20%)
[P31113]Overall time: 452.33 seconds
ubuntu@ip-172-31-51-147:/mnt/ref/cushaw$
```

Figure 5.43 *When CUSHAW2-GPU finish the sequence alignment, you can see 98.2% sequences are aligned. Same FASTQ file, compare with Bowtie result in Chapter 4, only 0.2% difference. Total processing time about 7.5 minutes, it is acceptable*

5.4 The Reason for the Limited Success of GPUs

The preceding section has illustrated the advantages of graphics cards in support-ing intensive computation, but a reader who has read articles on bioinformatics tools, will immediately realize that the use of traditional processors and software development on memory execution architecture are still mainstream. If the graph-ics card is really very good as a co-processor, and since CUDA is also very mature, in theory, there should not be such a dearth of software development. What is the reason?

Since CUDA is currently already mainstream, let us just analyse the situation for CUDA. Regardless of the hardware product or the CUDA version used in the latest standard, at this stage, the life cycle of the two is actually very short. New hardware emerged, and new software versions come out not long after, and imme-diately there is a new release ready to roll. This investment in resources is too expensive to keep up for most. For example, earlier products do not support the more precise double precision numerical calculation, which in scientific comput-ing is great influence on adoption. The performance of the hardware in keeping up with the times quickly, coupled with the software which has to be modified to fit the compiler architecture, wreaks havoc with end-users. Previously developed software versions are not necessarily just recompiled with minor modifications in order to leverage on the advantages of the hardware to achieve better performance. Much work needs to be done.

Maintenance costs are too high in proportion to the improvement in execution performance. So it is one thing to develop CUDA-based bioinformatics tools, but there is almost no support for backward compatibility with the old version. In contrast, research institutions have invested more in the CPU, which traditional tools rely on, presumably because of an inherited legacy programming framework. Moreover, the progress in processor speed, with improvements in the motherboard bus architecture which transfers data directly to the processor, performance is not necessarily worse than that of CUDA.

The "bus" is a hardware architecture terminology and I am not going to give too detailed description, especially on its rich history. I will just use the analogy of "transmission lines" to describe the instruction set. Simply put, CUDA is NVIDIA graphics cards working in the computing architecture. While the graphics card itself does not look like the central processor on the motherboard, but directly on it is reserved a slot for the GPU, and the direct use of the pipeline has the fastest memory communication. Results of the processor calculations can be quickly sent to the memory.

We should all know, that the speed of memory is a lot faster than the hard drive. Memory, is now very cheap, and acquiring a computer with 16 GB or higher memory capacity is not difficult. A middle-end server equipped with 128 GB or 256 GB memory is nothing special today. In contrast, the graphics card does not have direct communication with the memory on the motherboard, and the direct pipeline speed on the PCI-Express slot on the processor is far less than the direct communication with the memory. So one has to draw on the memory on the GPU card itself, the on-board memory. This memory, the Graphic Double Data Rate (GDDR) although fast, but the price is prohibitively high for the card to increase its capacity in this direction.

Currently for NVIDIA's latest CUDA-based Pascal products, the maximum capacity is no more than 12 GB (Titan X Series). If data exceeds this capacity, we need to tap conventional memory as a workaround. The graphics engine has a lot of kernel power, like a powerful generator with the limitation of external transmission of electricity due to the low carrying capacity, the data pipeline for the GPU is just too small; thus there is no way to have efficient delivery of the computing power.

The CUDA graphics card memory is also too small for the big datasets of bioinformatics computing, where large amount of data is required in genome sequence alignment. Such large data cannot be sent to a graphics card's memory, so through clever programming the information sent to the graphics card to leverage on its power in parallel computing and the results are then routed back to the hard disk, and then merged in the central processor.

So it can be seen that CUDA still has a high dependency on the traditional central processor. For example, how to split the data and route it between processor and memory, from the beginning, has been inseparable from the central processor. Currently released CUDA bioinformatics tools can be found with best performance figures for small datasets, such as E. coli bacterial genomes, but its ability to handle large genomic sequence sets is really not up to scratch.

"Compare with CPU-based alignment, is the alignment performance and result from CUDA acceptable?" We are interested in the answer, so we apply the same FASTQ dataset as input, hg19 is reference sequence. Benchmark the three sequence alignment tools; BWA, Bowtie2 and CUSHAW2-GPU.

BWA and Bowtie2 are CPU-based, we set the CPU cores parameter as 8. The results showed that Bowtie2 only required 5 minutes and 10 seconds, while BWA took 6 minutes and 50 seconds but CUSHAW2-GPU took 7 minutes and 32 seconds. On the surface, it seems the time difference between the above three is not significant. Actually, CUDA is more expensive. You have to invest more money for the CUDA card. With the same budget, you may prefer a higher specification processor. At least the CPU can be more widely used.

CHAPTER 6

Establishing a Research Workflow Pipeline

6.1 Automating Your Computational Workflow

The bioinformatics research process often involves multiple steps and many sets of tools. Typically, the original raw data file is sent to the first tool, and then the output is handed over to another tool to compute the results. Between these two steps, perhaps because of data format problems, the output of the first step may have to be processed by a third software tool, before it can be accepted by the next tool in the next step.

For example, to find a single point mutation in a study on Single Nucleotide Polymorphism (SNP) for a particular disease condition, one would start must sequence alignment, and then the results analysed with a gene-shaped? Verification or validation tools (such as Genotyper) to confirm whether it is a mutant. If you follow the traditional method of operation for beginners, each step has to be manually run with each tool, and then the result manually transferred to the next step using another set of necessary tools. This method works but it is not efficient enough for high-throughput.

Some bioinformatics computation can be very time consuming, and you may be able to predict when it will complete. If the completion time is early in the morning, you would not be pleased to have to get up to perform the next step, right? In fact, the biggest advantage of carrying out bioinformatics experiments and procedures is not having to wait up late for a wet lab experiment to complete! If you can program your computational work to completion, and get up in the morning to see the results, this would probably be the ideal situation.

Computational time is one thing. Sometimes the process may be so cumbersome that to complete an entire experiment, one has to perform more than 10 steps. Due to the complicated steps, one may make mistakes. For instance, while waiting for the next command prompt when the execution of a step gets completed, you may be jotting in a laboratory notebook which is the next step to carry out, and still having to remind yourself to pay attention to the workflow and provenance. But as long as the input passes through the human hands, there is potential for error. How can we create a complete experimental procedure, and even have it become a "standard operating procedure" (SOP), to standardize the provenance and ensure reproducibility of your work?

In this chapter, we will learn how to build a properly executable, complete series of computational steps, or "Macro", where the format definition of the input data is taken care of, and where we can await the results of the workflow. We will be giving you some tips on how to make your computing processes more readable and hence more easily maintained. We will also learn about potential errors you may encounter, so that your own design and computational processes can be more easily debugged.

One major goal of bioinformatics is to develop usable tools that is integrated with other research processes in the laboratory. A casual glance through publications on bioinformatics tools will show that many are built upon a process template or experimental protocol, and then released for others to use. If you are determined to develop tools with such functionality, it is important to learn at least a scripting language and how we can chain together in an explicit workflow, all individuals steps that make up the complete analysis.

6.2 Scripting Language

A scripting language (Script language), is defined in information science as a programming language which can create code, whose process is performed without needing to be compiled by the script parser, and can be performed directly after the direct interpretation of the language. Our aim is to make the individual separate Linux commands executed in sequence launched once to complete all the work.

The simplest way to do this is (although not the most efficient) is to record each command into a text file, then copy and paste each command into the terminal prompt and press Enter to execute. This is not really building a process pipeline. What is more efficient and what we wish to achieve, is to use the Linux commands as a Shell Script using the bash Shell Scripting language. This command shell facilitates a defined command terminal environment that is consistent and reproducible for a sequence of commands to be executed. Our learning goal is to complete the work in sequence, and then to allow the user to modify the process at least at the script parameter level, so as to facilitate the re-use of the program, or in some cases, to re-purpose the script to carry out another similar task.

To get this sequential work done is almost as simple as "copy and paste", so long as you have a complete record of each computing step, none of which can be executed from the beginning to the end, and not a series of actions which a development tool can completely perform. The Shell Script is a program designed to instruct the computer what to do for each step of your workflow sequence. Until now, you can copy and paste (or use the Ctrl arrows to recall the command history, or re-run or edit your commands). Using scripting language to run the commands, you can do it easily as long as you can figure out the order and sequence among each command. The aim of "modifying the least parameters" will depend on the characteristics of your analysis tools, and what needs to be adjusted where and when. Below is a basic Shell Script with elements which you just need to modify in a few places, and you can have a proper fully automated computational research process built into the script for direct execution or modification for re-use.

Before we embark on the details of the Shell Script, take the examples in the previous Chapter 4, as given below:

```
./bowtie-build ../UCSChg19/hg19.fa hg19

./bowtie  ../../ref/hg19  ../../data/SRR101437.fastq
-p 4 -S > test.sam

./bowtie  hg19  SRR101437.fastq  -p  4  -n  3  -S  >
test2.sam

sudo apt-get install samtools

./samtools view -bS ~/result/result.sam >
~/result/result.bam

./samtools sort ~/result/result.bam
~/result/result.sort

./samtools index ~/result/result.sort.bam
```

You can actually save each of these commands into a file, make it executable as explained in Chapter 2, and run the file as an executable one. The shell will open the file, take each line of the file and execute them as if you had actually typed it at the command prompt. In other words, if you have just executed 20 commands at the command prompt and save all these commands in a file made executable, you can "re-run" all these commands from the beginning. And typically if the input file is say from another sequencing run, and you need to repeat this exact same sequence of commands, all that needs to be done is to change the input file name and a few directory folder paths.

Script command

And instead of copying and pasting these commands into a file, you can actually record every command that you type into the command prompt by entering the command, `script`. By default, the script will save every command you enter, into a file called typescript in the current working directory. When you are done, simply enter the command, `exit`, to stop saving commands into your typescript file. Next, simply `mv typescript myscript.sh` and `chmod +x myscript.sh`, and preferably edit myscript.sh using a text editor, and check if all the commands are properly saved, and typographical errors are removed, and you will have a Shell Script file readily made for execution. (Note that all these commands you have keyed in, are actually also stored automatically in the file .bash_history without even having to use the script command. `tail -10 .bash_history` will produce the last 10 commands you have entered.) We will next show you the anatomy of a Shell Script file, and how you can parameterize the filenames and customize the script for re-use.

Element 1: Header

The Header in programming usually defines the main points of the program, telling the user what this program does and what language is used. It allows any programming language compiler or parser to know what language it is written in.

In Shell Scripting, the first line is usually `#!/bin/bash`. This means that the Shell interface, bash, is to be used to interpret the script following thereafter. Since the bash shell, which is also a program in itself, is usually placed in the folder /bin, the full path is specified after the `#!` (Fondly called "hex bang" or "she-bang".) You can edit this file using a text editor, which is analogous to editing Word files, except that it is in plain text. As a convention, we can save the file with an appropriate extension, such as .sh, which is the usual Shell Script extension. For me, I name my files as simply as I can in order to simplify typing.

Because there are many interpreted scripting languages, Perl, Python, Ruby, JavaScript, we can define the script interpreter where we just defined "bash", e.g. /usr/bin/perl or /usr/bin/python, etc. Typically by convention, they also have a specific extension, namely, .pl, .py, .rb, .js, just to make it conveniently obvious what kind of script they are.

Element 2: Variables

A variable in Shell Script is represented by a string of characters prefixed with $, e.g. $var2, $x, $new_string, are all valid variables. This is the way a computer program allows the value contained in the variable to be defined/assigned or changed/reassigned. It is a great way for parameterization, especially where the same parameter is re-used several times. This is done through the process of variable substitution at the place where the variable occurs.

Example: The set of commands used in Chapter 4 can now be parameterized by using variable names.

```
#!/bin/bash
./bowtie-build ../UCSChg19/hg19.fa hg19
bowtie ../../ref/hg19 ../../data/SRR101437.fastq -p
4 -S > test.sam
bowtie hg19 SRR101437.fastq -p 4 -n 3 -S > test2.sam
sudo apt-get install samtools
./samtools view -bS ~/result/result.sam >
~/result/result.bam
./samtools sort ~/result/result.bam
~/result/result.sort
./samtools index ~/result/result.sort.bam
```

Take a look at the commands used in Chapter 4. You can immediately spot some potential parameters which have been repeatedly typed into the command line. These can be assigned to a variable and then reused at the same locations using the variable in place of the parameter itself. Indeed, the use of a set of reference sequences is frequently the same, so you could parameterize the filename and the path.

In the study of the human genome, there is typically the UCSC HG19 dataset used as the reference sequence. The path may very long, so why type these characters repeatedly and why not specify a variable called hg19 and assign it with the path?

```
hg19="/home/eric/UCSChg19"
bowtie-build $hg19/hg19.fa hg19
```

So the shell interpreter will assign /home/eric/UCSChg19 as the value of $hg19, the variable. Wherever $hg is invoked, shell will interpolate the value and insert it at that location, in a process called variable substitution, before executing the command:

```
bowtie-build $hg19/hg19.fa hg19
```

becomes

```
bowtie-build /home/eric/UCSChg19/hg19.fa hg19
```

Note that $hg19 is the variable and hg19 is the string telling bowtie-build to build the database index into it.

For instance, the sequence alignment tool BWA is kept in my home directory /home/eric. Inside this home folder, is refidx/bawIdx, the reference sequence classification. The reference sequence is called ucsc.hg19.fasta.

The full path is thus /home/eric/refidx/bwaIdx/ucsc.hg19.fasta. This name is very long and bits of which may be reused. So we can assign a variable using:

```
hg19="/home/eric/refidx/bwaIdx/ucsc.hg19.fasta"
```

(This reference sequence may be used by the sequence alignment software bowtie, to build a series of index files, e.g. ucsc.hg19.fasta.bwt1, ucsc.hg19.fasta.bwt2, etc., automatically using the first part of the file name in front, to generate a series of reference sequence index file names. Therefore, pay attention to the path variable in the design of your Shell Script.)

Which variables could be used based on rarely changed parameters. Another reason is to avoid too many things to change. So the path is often long in the code path definition. If you do not use variables and directly have the entire paths "hard-coded" in the script program, then it will be very troublesome to update or maintain the script because every time you want to change a dataset, you have to change the path at multiple locations in the code. By using a variable definition, and using a variable name as the placeholder, where shell will expand the value during variable substitution, then we merely need to make the update at a single location, at the variable definition line. In this way, we avoid having too many things to change.

Another parameter that could lend itself to be assigned as a variable, is the parameters used by the program itself. For example, sequence alignment can be adjusted to allow for less than a certain Mismatch number. So we can assign `Mismatch=2` in a shell variable statement, and thereafter quote `$Mismatch` to mean the value 2.

In short, the variable naming process can be very flexible, in that anyone can decide where to assign a suitable variable and what to call it. Therefore it is best to establish a good naming convention. In the IT industry, professional programmers typically use a "Hungarian notation" (this is named after a Microsoft programmer called Charles Simonyi who invented it, Simonyi is Hungarian). Usually the first word will be lowercase, and then followed by the uppercase second word, e.g. "setMismatch".

Placed inside the original script program, they look like this:

```
bowtie $hg19 .. / ../data/SRR101437.fastq -p 4 -n
$setMismatch -S > SRR101437.sam
```

By comparing with the original command,

```
bowtie ../../ref/hg19 ../../data/SRR101437.fastq -p
4 -n 2 -S > SRR101437.sam
```

You can see where the variable name gets substituted.

In the example above, the reference sequence is `$hg19`, and the sequence alignment is less than the allowed number of `$setMismatch`. Thus a variable in Shell Scripting can be assigned a string or a number.

You may be clever enough to note that we can also use variables for the number of processor cores and even the SaveAs results file name can be assigned as variables. Indeed, these are possible. One additional tip: variable names can be combined to create a longer string when being interpolated by shell. For example, in a series of experiments, filenames may be called SRR101437. We can assign `$defSeries` to the specified value, `SRR101437`, written as: `defSeries=SRR101437`.

In the Bowtie example above, the last part can be changed to $defSeries.sam, and the output will be saved as a new file called SRR101437.sam. If you have a series of files, SRR101437..438 etc., you may want to create two variables instead, defS="SRR101" and num=437 and use $defS$num.sam for shell to interpolate into SRR101437. In this way by changing 437 to 438, etc., you can reuse the program again and again. Later on we will show you how you can pass a command line argument into a variable inside the program using $1 $2, etc., for matching each argument stated on the command line.

Element 3: Comment

There is an important principle in programming, and that is a code should be readable. A program that is impossible to read, is so obscure that probably nobody, not even the programmer who wrote the program himself, can maintain or debug it. To make the code more readable, we should have annotations to explain the code. This is achieved by something called the comment line, which as the character # preceding the comment until the end of the line.

For example, all we need to do is to simply prefix the line with # and we can make notes, for example, at the beginning of the program, to indicate the original creation date of the program, the developer name and contact, the program's main function. Because these annotations are prefixed by a comment marker #, they will not be executed; the shell interpreter and parser will skip these lines.

Many programmers use the comment lines to insert remarks about known bugs or programming issues and other annotations, thus allowing other subsequent programmers who are revising the code to add their own comments as well. This helps users reading the program code to appreciate the development of the program. Even if the program is written for your own private use, it is good practice to insert your own annotations. After a period of time, we may open the source file, and realize that we have forgotten what the program does and how we achieved the results. Strategically located, and well-written comments and annotations will rekindle our memory.

Shell Script annotations use the # sign. For example, to illustrate the significance of variables:

```
#Define the mismatch numbers
setMismatch=2
```

Alternatively,

```
setMismatch=2 #insert the mismatch number here
```

Anyone who wants to modify this program and re-use it for a different mismatch threshold, can find this location and edit a different number here.

To prevent any encoding issues, try to comment using standard English to avoid any ambiguity or code distortions. All successful implementations of software code, almost invariably have well documented and annotated code using the comment lines.

If you have written HTML code for web pages or written in any other programming language, you would ask how about block comments, or comments which stretch more than one line. In Shell Scripting, the block comment syntax is not provided, so every comment line has to be prefixed with the # sign.

Element 4: Passing variables from the argument line

Suppose the command for the above script is

```
./myscript.sh
```

We can actually parameterize $setMismatch to read the number from the command line.

Inside the code,

```
setMismatch=$1
# read $1 from the command line first argument
./myscript.sh 3
```

will actually pass the number three to $1, and shell interpolation will carry out variable substitution at that line in the code to execute setMismatch=3.

In this way, we could run a series of complex commands, using different numbers

```
./myscript.sh 1
./myscript.sh 2
./myscript.sh 3
```

And so on, which can get pretty tedious. The lazy ones amongst us will surely complain that there must be a simpler way of doing this, and indeed, there is. We can use the shell iterator

```
for var in 1..5
do
./myscript.sh $var
done
```

and all this can be squeezed into the command line:

```
for x in {1..5}; do ./myscript.sh $x; done
```

This will essentially set the variable $setMismatch to $1, which in turn was iterated from 1 to 5.

1..5 is called the range and the braces {} carry out the brace expansion of the range and replaces with 1 2 3 4 5 at the location where {1..5} occurs. In this way, the "for x in" command will be executed by shell as for x in 1 2 3 4 5, which does the iterations of whatever occurs after the "do" five times.

This iterator is also called the bash for loop, which loops through whatever occurs between the do and the done. Multiple commands can be separated by a semicolon (;).

This iterator is a very powerful programming idiom. It is worth our while to look at it carefully, when we talk about "command substitution", which is analogous to "variable substitution" which we have learnt earlier.

For example, if we want to iterate SRR101002 to SRR1012, we make use of the brace expansion but the older bash version, {02..12} will produce 2, 3, 4, 5 ... 12. What we need is to have the single digits padded with a zero. The UNIX command seq -w comes in handy.

```
seq -w 2 12
02
03
04
05
...
11
12
```

So if we perform command substitution as follows:

```
for i in $(seq -w 2 12); do echo SRR1010$i; done
SRR101002
SRR101003
SRR101004
...
SRR101011
SRR101012
```

This helpful little code snippet illustrates the power of using a command inside another command, that is, the use of command substitution $( ). Alternative notation is the backtick `. `seq -w 2 12` will perform command substitution for older bash versions.

Element 5: Forced line breaks

Some bioinformatics tools require many user-supplied arguments. Some of these arguments are very long. Often they stretch over one line, and overflow into the next, depending on your window size. To avoid such an error-prone situation, we recommend the practice of forced line breaks to improve visibility. Just enter a backslash (\) as a continuation line indicator, and then press Enter to enter the rest of command.

For example, we could split the long Bowtie command, which stretches two lines, into multiple lines simply by ending each line with a backslash. This conveys to the shell interpreter that the end of the line has not been reached yet, and to continue looking for more parameters in the command.

```
bowtie $hg19\
../../data/SRR101437.fastq\
-p 4\
-n $setMismatch\
-S > SRR101437.sam
```

In the process of aligning sequences, we can use forced line breaks and variable substitution or command substitution to make the code look cleaner, and readable. The use of this kind of layout style is a good practice to acquire in programming. In addition, it helps in maintainability of the code, and reduces the possibility of keyboard input error.

6.3 Testing and Debugging

Every program, no matter who or how good the programmer is, needs to go through a testing procedure. The main purpose is to test if the program execution with test datasets will produce the anticipated results in line with expectations. The earlier errors are detected, and quickly corrected, the better it is. This will avoid the debugging predicament where one wrong step triggers a series of wrong steps, especially when the program is a sizeable one.

In bioinformatics workflows and pipelines, Shell Scripts debugging is straightforward and essential as well. Often, the computational procedure is developed by others, and so are the tools used. Hence assuming that the programs are without bugs, the results will still have to be checked and the parameters determined to be correct. Because of the large amount of data generally fed into a bioinformatics workflow, if the parameter settings are incorrect, the waste of computing resources and disk space storage is not worth it. This section will explain how to perform efficient testing and debugging.

Keeping track of the current project

The biggest fear in programming is that you do not know when something has gone wrong. For Shell Scripts used in workflows involving bioinformatics tools used sequentially, one common problem is when one of the tools has finished running, the next tool has no way to pick up the output as its input. Subsequent programs cannot be executed because of this failed step, and a whole string of error messages are generated. This can be systematically traced from where things started to go wrong, but it is very tedious. Therefore, the easiest way to identify when each step of the program is successfully completed is by using the echo command to "echo" a statement to the standard output. In this way, we can carry out custom work and keep track of where the running Shell Script is stuck at.

The meaning of echo as a Linux command is to print a specific message to the standard output. In the past, the terminal host connection was very slow, and the purpose of "echo" is to identify whether the host key has been received.

To output a complete string, the simple command is:

```
echo "string"
```

For example, enter

```
echo "Bonjour"
```

After pressing Enter, the word "Bonjour" should be displayed immediately. So we can use this command to display the UNIX command in the Shell Script that is currently running. For example, the work being performed is sequence alignment, and in order to track the progress of the program with a visually prominent output flanking where the script starts and stops.

```
echo "=====Alignment Process Start====="
```

```
echo "=====Alignment Process End====="
```

Thus, these flanking messages will indicate to us when the program has reached the start and when it reached the end. So if there are errors in the script, they willl appear between these two marker prompts. Sometimes adding another blank line with a newline character (\n) will also help increase the visibility of output, and enable us to detect any problems.

For example, the following code

```
echo "=====Alignment Process Start=====\n"
```

This improves the clarity of the screen output when the program is running, to have the report of that echo command to be clearly distinguished from other lines of output messages that are also displayed when the program is running.

Complementing tests of code blocks

The previous method of reporting the progress of the program will roughly guide you to where the errors are located in the code. To confirm the exact location of any error, re-run the entire process using the partial execution approach. Let us explain this approach of selectively commenting off segments of the code to narrow down where the bugs are located.

In the previous section, we have talked about the annotation feature that allows the source code in some places not to be executed. You can now use this commenting function, to comment off the processes which you do not want to be executed, by prefixing them with the # sign. This approach is tedious and requires some effort, in order to correlate for each process the relationship between the input command and the output, in order to identify where the bug is located.

For example, if you guess a certain line of code is causing the trouble, you can comment it off and if the error disappears, then you will know that the trouble is generated at that line of code which was commented off. Once identified, you will have to analyse why that line of code is giving the error.

Perhaps the error occurs because the output file name is set incorrectly, resulting in the next step not being able to find this file name and not being able to feed into the next step of the pipeline process. For example, some bioinformatics tools will output the results into user-specified files names other than its own default, but it also adds the file extension. So when you created the script that adds an extension explicitly, you may end up with two extensions. This will obviously cause problems in the next step.

For example, Bowtie indexing automatically generates an output file with the .bai extension, but if you accidentally wrote the command in the form where you specified a file name with .bai, Bowtie may add another extension resulting in a filename with .bai.bai as a double extension. Do be careful to avoid such errors.

Sometimes there is a situation, where we are doing the block testing by commenting off the respective lines, we accidentally add too many or too few #. This will generate extra or missing output, resulting in an incorrect block test. This human error can be prevented by taking extra care during the editing of the code, or by using a better program text editor, which will allow you to see the commented block clearly, and allow you to identify any errors immediately. Similarly, it helps to be able to see clearly when you start to uncomment the code blocks when the troubleshooting is completed.

In addition, it helps also to alert you during the production of the final version of the code, to delete some of these commented lines and other annotations which you have added to help you test the code, so that the ultimate program looks a little cleaner and less cluttered.

Calculating the execution time

When bioinformatics tools perform computational analysis, sometimes they take a lot of time. You may be queried, "How long is this process going to take?". Obviously you cannot stay in front of a computer with a stopwatch. Moreover, if you need to publish papers describing the tools you have developed, you also need to explain the execution time or show performance analysis results. Hence, we need a more precise figure and a more effective method of knowing this. In order to do this, we can use the time feature of the Linux operating system, which can give you an accurate computation of the total time elapsed for a computation.

Here is a sample code where you can insert the time stamps of when the shell executes the beginning and the end of the code, and thereby giving you the ability to compute the total run time. Here is a code example:

```
#!/bin/bash
START=$(date +%s)

#Insert here the commands you wish to run
END=$(date +%s)

DIFF=$(( $END - $START ))

echo "It took $DIFF seconds"
```

For six lines of code, this is a very simple and effective process. Let us take a look at how this works. At the beginning, we defined the START variable by using command substitution, i.e. $( ) syntax. Here the UNIX command is the date command, followed by the +%s argument. From the manual page of the date, you will realize that this expresses the computer OS current time in seconds and stores it in the variable $START. When the shell reaches the assignment of the END variable, it does the same thing at a later time, almost stamped in seconds with the +%s argument. Next, using a variable named DIFF, and the $(( )) syntax, which is the bash syntax for carrying out a mathematical evaluation of the expression inside the double parentheses, we will be able to subtract the time in seconds at the start from the time in seconds at the end, thereby producing the total time elapsed in seconds as the final output value in the next command, echo.

We used seconds as the unit to consider the speed of the execution, but frequently we will get a better feel of the speed if we divide the time by 60 or 360 to get the time minutes or hours, respectively.

```
minDIFF=$(( $END - $START / 60 ))

hrDIFF=$(( $minDIFF / 60 ))

echo "It took $DIFF seconds or $minDIFF minutes or
$hrDIFF hours"
```

Now there is also another method that does not require you to embed "date" commands in the code. It is to use the time command instead. So if you need the running time of a program, prefix the command with the "time" command. This will allow the command to run as normal, but keep track of time, and then print out the time elapsed after the entire output is completed. Sometimes to avoid the output obscuring the time stamps, we can pipe the output to a black hole /dev/null or to a output file, using the Unix redirection >. For example, if we have a script called mysuperscript.sh

```
time mysuperscript.sh > /dev/null

real 0m5.462s

user 0m5.432s

sys 0m0.012s
```

This program took 5.462 seconds of wall clock time (real) from start to finish of the program, the actual elapsed time. The user time refers to the amount of CPU time spent in processing the user code within the user's process. Other processes and the time the process incurs when blocked, does not count towards this figure. The sys time refers to the amount of CPU time spent in system calls within the Linux kernel, as opposed to the user run library code. user + Sys time will tell you how much CPU time you have used, and if there are more than one processor being used by multiple threads of your process, then this time may exceed the wall clock real time.

6.4 Implementation Case Studies

The following code is a simple example of a program I made for a small project. It automatically downloads all the bioinformatics tools needed, and decompresses compressed files, and compiles the source code and finally saves the binary executables thus compiled to their designated folder. We can assess the quality of my code, although not the best, but the basic aim is achieved.

```
#!/usr/bin/bash
#Prepare the tools for my brain tumor research pieline.
#Ver 0.1
#Eric C.Y. LEE 2011.Dec
###Download the tools
##Samtools
echo "====Start to download Samtools latest version===="

curl -O -L http://downloads.sourceforge.net/project/
samtools/samtools/0.1.18/samtools-0.1.18.tar.bz2

tar -xvjpf samtools-0.1.18.tar.bz2

mv samtools-0.1.18 samtools

rm samtools-0.1.18.tar.bz2

cd samtools

make

cd ..

echo "Samtools latest version download and setting
complete!\n"

##BWA
echo "====Start to download BWA latest version===="
```

```
curl -O -L http://downloads.sourceforge.net/project/
bio-bwa/bwa-0.6.1.tar.bz2

tar -xvjpf bwa-0.6.1.tar.bz2

mv bwa-0.6.1 bwa

rm bwa-0.6.1.tar.bz2

cd bwa

make

cd .. echo "BWA latest version download and setting
complete!\n"

##GATK
echo "====Start to download GATK latest version===="

curl   -O   ftp://ftp.broadinstitute.org/pub/gsa/
GenomeAnalysisTK/GenomeAnalysisTK-latest.tar.bz2

tar -xvjpf GenomeAnalysisTK-latest.tar.bz2

mv GenomeAnalysisTK-1.3-24-gc8b1c92 gatk

rm GenomeAnalysisTK-latest.tar.bz2

echo "GATK latest version download and setting
complete!\n"

##Picard
echo "====Start to download Picard latest version===="

curl -O -L http://downloads.sourceforge.net/project/
picard/picard-tools/1.57/picard-tools-1.57.zip

unzip picard-tools-1.57.zip

mv picard-tools-1.57 picard

rm picard-tools-1.57.zip

echo "Picard latest version download and setting
complete!\n"

##snpEFF
echo "====Start to download snpEFF latest version===="

curl -O -L http://downloads.sourceforge.net/project/
snpeff/snpEff_v2_0_4_rc1_core.zip
```

```
unzip snpEff_v2_0_4_rc1_core.zip

rm snpEff_v2_0_4_rc1_core.zip

mv snpEff_v2_0_4_rc1 snpEff

echo "snpEFF latest version download and setting
complete!\n"

###Download the reference data ##ucsc.hg.19 mkdir refidx
echo "====Start to download ucsc HG.19 related data
and index files.===="

curl -O ftp://gsapubftp-anonymous@ftp.broadinstitute.
org/bundle/1.2/hg19/ucsc.hg19.dict.gz

mv ucsc.hg19.dict.gz refidx/gunzip
refidx/ucsc.hg19.dict.gz

curl -O ftp://gsapubftp-anonymous@ftp.broadinstitute.
org/bundle/1.2/hg19/ucsc.hg19.fasta.fai.gz

mv ucsc.hg19.fasta.fai.gz refidx/

gunzip refidx/ucsc.hg19.fasta.fai.gz

curl -O ftp://gsapubftp-anonymous@ftp.broadinstitute.
org/bundle/1.2/hg19/ucsc.hg19.fasta.gz

mv ucsc.hg19.fasta.gz refidx/ gunzip refidx/ucsc.hg19.
fasta.gz

curl -O ftp://gsapubftp-anonymous@ftp.broadinstitute.
org/bundle/1.2/hg19/ucsc.hg19.stats.gz

mv ucsc.hg19.stats.gz refidx/

gunzip refidx/ucsc.hg19.stats.gz

echo "ucsc HG.19 related files extract completely!\n"
##dbSNP
mkdir dbsnp

echo "====Start to download dbSNP_132 related data
and index files.===="

curl -O ftp://gsapubftp-anonymous@ftp.broadinstitute.
org/bundle/1.2/hg19/dbsnp_132.hg19.vcf.gz
```

```
mv dbsnp_132.hg19.vcf.gz dbsnp/

gunzip dbsnp/dbsnp_132.hg19.vcf.gz

curl -O ftp://gsapubftp-anonymous@ftp.broadinstitute.
org/bundle/1.2/hg19/dbsnp_132.hg19.vcf.idx.gz

mv dbsnp_132.hg19.vcf.idx.gz dbsnp/

gunzip dbsnp/dbsnp_132.hg19.vcf.idx.gz

echo "dbSNP_132 related files extract completely!\n"

##snpEFF database
echo "====Start to download snpEFF database file.===="

curl  -O  -L  http://sourceforge.net/projects/snpeff/
files/databases/v2_0_4/snpEff_v2_0_4_hg37.63.zip

unzip snpEff_v2_0_4_hg37.63.zip

rm snpEff_v2_0_4_hg37.63.zip

mv data snpData

echo "snpEFF database file extract completely!\n"
```

The example above contains basic annotations to echo the instructions to download the bioinformatics tools. The Shell Script is considered very simple, and does not use variable definitions. The script, though reusable, is meant just for a time-saving tool for downloading and compiling hard-coded file names of software packages with explicitly stated version numbers. I admit that it could have been better, for example, if the URL of the bioinformatics tools have been assembled at the beginning and assigned to variables by variable definition statements. The next script is slightly longer, and the tools have more parameters.

```
#!/usr/bin/bash
#My bioinformatics pipeline which for brain tumor research.
#Ver 0.1
#Eric C.Y. LEE 2011.Dec
###Path Definition
##Tools
#GATK
GATK=/Users/Eric/exam/gatk/GenomeAnalysisTK.jar
```

```
#Picard
PICARD=/Users/Eric/exam/picard/

#snpEFF
SNPEFF=/Users/Eric/exam/snpEff/snpEff.jar

#snpEff config
SNPEFFCOFING=/Users/Eric/exam/snpEff/snpEff.config

##Data
#Reference Genome
hg19=/Users/Eric/exam/bwaidx/ucsc.hg19.fasta

#SNPs
dbsnp=/Users/Eric/exam/dbsnp/dbsnp_132.hg19.vcf

#VCFs
vcf=/Users/Eric/exam/dbsnp/

#Run files
SRR=/Users/Eric/exam/SRR101437.fastq
SRRNO=101437

####Start process
echo "====alignment===="

#alignment
bwa aln -t 8 $hg19 $SRR > SRR$SRRNO.sai

#convert to sam
bwa samse $hg19 SRR$SRRNO.sai $SRR > SRR$SRRNO.sam

#convert to bam
samtools view -bS SRR$SRRNO.sam > SRR$SRRNO.bam

#sort and index
samtools sort SRR$SRRNO.bam SRR$SRRNO.sorted
samtools index SRR$SRRNO.sorted.bam
```

```
#Reorder the SAM
echo "====Reorder SAM===="

#Sort the SAM
java -jar -Xmx4g -Dsnappy.loader.verbosity=true\
      $PICARD/SortSam.jar I=SRR$SRRNO.sorted.bam \
      O=SRR$SRRNO.resultCoordinate.bam SO=coordinate
      VALIDATION_STRINGENCY=SILENT

echo "====Remove duplicate===="
#Remove the duplicate
java -Xmx4g -jar $PICARD/MarkDuplicates.jar
      I=SRR$SRRNO.resultCoordinate.bam \
      O=SRR$SRRNO.resultCoordinateRmdup.bam
      METRICS_FILE=met.file
      REMOVE_DUPLICATES=true

samtools index SRR$SRRNO.resultCoordinateRmdup.bam

echo "====Count covariates===="
#count covariates
java -Xmx4g -jar $GATK\
      -R $hg19\
      -knownSites $dbsnp\
      -I SRR$SRRNO.resultCoordinateRmdup.bam\
      -T CountCovariates -cov ReadGroupCovariate
      -cov QualityScoreCovariate\
      -cov CycleCovariate -cov DinucCovariate\
      --default_platform illumina
      --default_read_group SRR$SRRNO\
      -recalFile SRR$SRRNO.recal_data.csv

echo "====TableRecalibration===="
#TableRecalibration
java -Xmx4g -jar $GATK -T TableRecalibration\
      -R $hg19\
      -I SRR$SRRNO.resultCoordinateRmdup.bam\
      -recalFile SRR$SRRNO.recal_data.csv\
      -o recalibrated.bam --default_platform Illumina\
      --default_read_group SRR$SRRNO
```

```
samtools index recalibrated.bam

echo "====Local realignment around Indels===="
#####Local realignment around Indels
#Creating Intervals
java -Xmx4g -jar $GATK\
      -R $hg19\
      -known $dbsnp\
      -I recalibrated.bam -T RealignerTargetCreator\
      -o SRR$SRRNO.forRealignerInterval.list

#Realigning
java -Xmx4g -jar $GATK\
      -T IndelRealigner -R $hg19\
      -known $dbsnp -I recalibrated.bam\
      -targetIntervals SRR$SRRNO.forRealignerInterval.list\
      -o SRR$SRRNO.RecalRealn.bam

#ReplaceReadGroup
java -jar $PICARD/AddOrReplaceReadGroups.jar\
      I=SRR$SRRNO.RecalRealn.bam
      O=SRR$SRRNO.RecalRealnRPG.bam\
      SORT_ORDER=coordinate RGID=SRR$SRRNO RGLB=bar
      RGPL=illumina RGSM=h19\
      RGPU=$SRRNO VALIDATION_STRINGENCY=SILENT

#Index SRR111940.RecalRealnRPG.bam
samtools index SRR$SRRNO.RecalRealnRPG.bam

#SNP Calling: UnifiedGenotyper
java -Xmx4g -jar $GATK\
      -T UnifiedGenotyper -R $hg19\
      --dbsnp $dbsnp -I SRR$SRRNO.RecalRealnRPG.bam\
      -o SRR$SRRNO.snps.raw.vcf -stand_call_conf 30.0

echo "====Variant quality score recalibration===="
#####Variant quality score recalibration
#VariantRecalibrator
java -Xmx4g -jar $GATK\
      -T VariantRecalibrator \
```

```
        -R $hg19\
        -input SRR$SRRNO.snps.raw.vcf \
        -resource:hapmap,known=false,training=true,
        truth=true,prior=15.0
        $vcf/hapmap_3.3.hg19.sites.vcf \
        -resource:omni,known=false,training=true,
        truth=false,prior=12.0
        $vcf/1000G_omni2.5.hg19.sites.vcf \
        -resource:dbsnp,known=true,training=false,
        truth=false,prior=8.0
        $vcf/dbsnp_132.hg19.vcf \
        -an QD -an HaplotypeScore -an MQRankSum
        -an ReadPosRankSum -an FS -an MQ \
        --maxGaussians 6 \
        -recalFile SRR$SRRNO.recal \
        -tranchesFile SRR$SRRNO.tranches \
        -rscriptFile SRR$SRRNO.plots.R

#ApplyRecalibration
java -Xmx3g -jar $GATK\
        -T ApplyRecalibration \
        -R $hg19\
        -input SRR$SRRNO.snps.raw.vcf \
        --ts_filter_level 99.0 \
        -tranchesFile SRR$SRRNO.tranches \
        -recalFile SRR$SRRNO.recal \
        -o SRR$SRRNO.recalibrated.filtered.vcf

####snpEFF
java -Xmx4G -jar $SNPEFF\
eff -v -i vcf -o vcf -c $SNPEFFCOFING\
hg37.63 SRR$SRRNO.recalibrated.filtered.vcf >
SRR$SRRNO.snpEff_output.vcf
```

In this second example of a Shell Script, I have started to define the main utility of the path, the path of the reference sequence, followed by comments and echo command to describe what kind of computation is being carried out.

At each step, the resulting output filename is also specified at the beginning of variable definition lines. Each result file name also facilitates continuity to the input of the next step. Each tool requires many parameters, therefore in order to look readable, I have included forced line breaks to split the complex parameters into independent lines.

6.5 Case Study of Common Mistakes

Finally, I want to analyse some of the common mistakes which beginners are prone to make. If you are building a bioinformatics research process, and there is an error, it may be from several sources, not just one. And you can also adopt several possible approaches for debugging. Here we share the design principles of the Shell Script taking into account the research needs and cross-platform issues.

Mistake 1: Confusing mess of relative paths

In the beginning of the chapter, I talked about the importance of the path on Linux. In particular "." (current directory) and ".." (parent directory one level above). Often, getting this confused is the source of many problems in scripting. Here are some examples in a program code excerpted to illustrate the problem:

```
###Download the tools
##Samtools
echo "Start to download Samtools latest version"
curl -O -L http://downloads.sourceforge.net/project/
samtools/samtools/0.1.18/samtools-0.1.18.tar.bz2

tar -xvjpf samtools-0.1.18.tar.bz2
mv samtools-0.1.18 samtools
rm samtools-0.1.18.tar.bz2
cd samtools
make
cd ..
echo "Samtools latest version download and setting
complete!\n"
```

In operations involving a change to another folder file name using mv, or using change directory (cd), one has to be careful. The original folder samtools-0.1.18 was changed to samtools and for compilation, we need to switch into this folder. After the compilation is completed, we should not forget to change the directory back to the parent level, using the cd .., to where the original work folder is. Otherwise, follow-up commands would have been inside this subfolder called samtools, the wrong folder. The occurrence of some errors is inevitable, if we do not take care to monitor what the current working directory is.

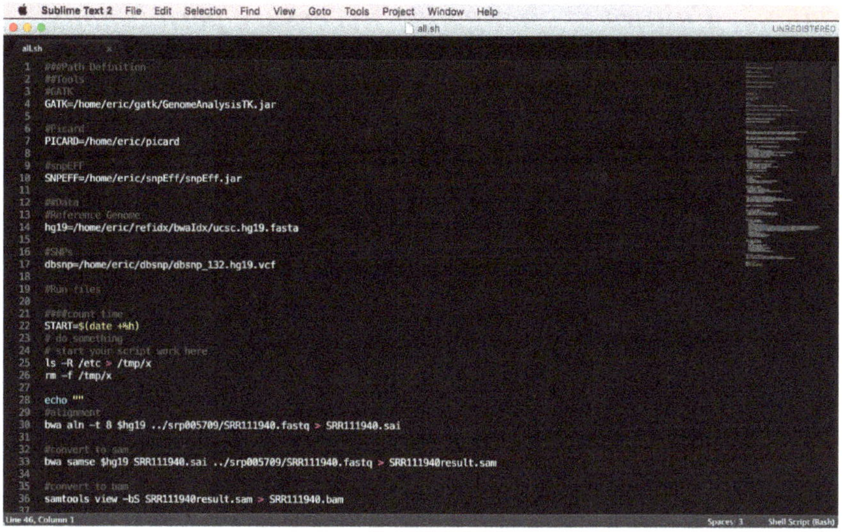

Figure 6.1 *A good text editor, especially a program editor, can reduce the chance of incurring errors. Such an editor will have colour-coded functions, with different colours denoting programming language annotations, variables, code and other entities, and overall, they generally improve programming efficiency. Pictured above is an example of a cross-platform text editor, Sublime Text*

Another mistake is that the folder does not exist or wrong, for example, in the following script to download and store sample of a portion of the reference sequence:

```
###Download the reference data
##ucsc.hg.19
mkdir refidx
echo "Start to download ucsc HG.19 related data and
index files."
curl -O ftp://gsapubftp-anonymous@ftp.broadinstitute.
org/bundle/1.2/hg19/ucsc.hg19.dict.gz
mv ucsc.hg19.dict.gz refidx/
gunzip refidx/ucsc.hg19.dict.gz
```

This folder, refidx, is where the downloaded reference sequence is stored. Start by creating this folder, using `mkdir refidx`, and then use `gunzip` to decompress the compressed reference sequence file with "absolute path" fully defined. Here we avoid switching to the refidx folder and having to change back to the working directory by `cd ...` Here the decompression is completed using the full path `gunzip refidx/ucsc.hg19.dict.gz`.

Mistake 2: Failure to change the necessary permissions

In the earlier chapters, we introduced the basic Linux command, chmod +x of files which we intend to make executable, by changing the UNIX permissions before trying to execute it. However, sometimes, even after doing this, we still get errors because in the implementation of some bioinformatics tools, administrator privileges are required, or during the computation process, a command is called which you do not have permission to execute. This latter problem is common, especially when a different user is called to execute the program, e.g. when a web server running under a user who has limited privileges is using CGI to call a script and execute it.

To prevent this problem from arising, there are two approaches. One is to log in as the administrator (root or superuser) and then execute the desired Shell Script. Another is to use temporary administrator rights to perform the command, which is frequently used in Ubuntu. If you do not have access to the root account (especially for security reasons), you will need to use the local administrator privileges by invoking the command prefixed with the sudo command. After pressing Enter, you will be asked to enter your password. By default, your installation of Ubuntu should allow you to use sudo ./example.sh. After you authenticate with your password, this command will be executed with administrator privileges.

To see if you can use the sudo command, i.e. whether you are a sudoer (superuser doer), you can check the file /etc/sudoers. If your userid is not present there, ask your server administrator to add you into the sudoers file. If you are a superuser, you can login as root, or su root, key in the password, and use the visudo command to safely edit the /etc/sudoers file to add your userid.

Mistake 3: The disk becomes full during execution

Bioinformatics computation, in many cases, involves multiple user shared hardware resources. Your hard disk is a resource which can be consumed very quickly, as long as users do not carry out regular housekeeping and pay attention to the organization of their files, the host's hard disk can be quickly filled. Once the hard disk is full, we can only resolve this matter through the system administrator's intervention. We have already introduced to you earlier the df command to check the current disk usage of your server, and the du command to calculate the disk usage of your files and folders. Since some bioinformatics tools can result in huge files and since there is no way to estimate in advance the final size of the files, we have to use our common sense and general experience to judge whether there is sufficient disk space left for running your bioinformatics tools. In particular, when decompressing the input sequence file, we generate another larger file and end up with more than twice the size of the file we originally downloaded. So wherever possible, delete temporarily created files, such as those created during compilation of a source code.

To solve the disk space problem, one must be conscious of files created and their sizes, and one must take steps to delete all unnecessary files once they are used. The second example above illustrates this, where the output file of one step becomes the input to the next step. Programming skills here help save space by deleting unnecessary files, some of which are created on the fly. The following examples illustrate how when execution of sequence alignment begins, and when sam and bam file conversion take place:

```
#alignment
bwa aln -t 8 $hg19 $SRR > SRR$SRRNO.sai
#convert to sam
bwa samse $hg19 SRR$SRRNO.sai $SRR > SRR$SRRNO.sam
#convert to bam
samtools view -bS SRR$SRRNO.sam > SRR$SRRNO.bam
```

The above process will output SRR$SRRNO.sai, SRR$SRRNO.sam with SRR$SRRNO.bam, a total of three files, but in the end the only one to be used is SRR$SRRNO.bam. What we can do under situations of disk space limitations, is to complete each step and then immediately delete the files generated in the previous step. Just make sure that the deleted file will no longer be required in the following steps. To save disk space, the code above after modification becomes:

```
#alignment
bwa aln -t 8 $hg19 $SRR > SRR$SRRNO.sai
#convert to sam
bwa samse $hg19 SRR$SRRNO.sai $SRR > SRR$SRRNO.sam
rm SRR$SRRNO.sai
#convert to bam
samtools view -bS SRR$SRRNO.sam > SRR$SRRNO.bam
rm SRR$SRRNO.sam
```

Because the disk full problem created a lot of unexpected interruption and unpredictable problems, including bad consequences, such as when editing a file half-way and the disk full problem triggers, the current file cannot be saved, and the program crashes, and the file data may be lost. This may not necessarily be your fault as it could be due to other users who temporarily generate such huge files that the disk space becomes full, and others in the middle of their program execution end up with data loss because after opening a file, they cannot write it back.

So in a situation where hard disk is cheap, do prepare for a large disk space or carry out housekeeping to keep a little more disk space as buffer just in case, combined with a conscious effort to delete temporary or intermediate files in a timely manner using the above techniques.

Mistake 4: Ignoring cross-platform shell portability considerations

Linux is a branch of the UNIX family, so is macOS. Although most of the commands are the same, and the look-and-feel is similar when we log in using ssh, we may not always pay sufficient attention to what kind of operating system is used by the host. Most bash shell commands are portable across platforms. However, we need to pay attention to environmental variables, as well as library paths, and other defaults which differ from platform to platform, Linux variants, etc.

So if you have just developed a Shell Script for your research workflows, and you wish to run it on another host machine or share it with other users, the program may not work properly due to various cross-platform compatibility issues. Take the example of the previous section, where to download the file, we used the wget command with the URL web address as argument, and now it became the curl command? The reason is because the wget command is not native to the macOS platform. It is a software which we have to install.

We have no way to guarantee that all macOS users will have wget pre-installed. For portability considerations, your best insurance is to use the curl command, which is common to both Linux and macOS. It has slightly different file download parameters with the -O parameter.

If you are designing bioinformatics tools, to openly share with others, you should take into account these cross-platform issues. To avoid this problem, have a variety of Linux platforms, macOS, FreeBSD (macOS main core is the FreeBSD Linux branch changes), etc., undergo the above tests, or add commands in your Shell Script to test for the presence of commands before executing them, and trigger user-friendly error messages when the program cannot find the required files or executables.

Sometimes the programs are there, but they are installed in a folder which is not in the search path $PATH of the user executing the program. You may want to include commented annotations at the outset of your program to warn users if they are using it on an unsupported platform, which you have not tested in advance.

In typically robust Shell Script with some degree of portability taken into consideration, we have to use the if-then-else-fi construct to test if the program exists.

Consider this code snippet:

```
URL="http://bowtie.sourceforge.net/bowtie.tar.gz"
fW=$( which wget )
fC=$( which curl )
if test -x "$fW"
then
        $fW $URL
elif test -x "$fC"
then
        $fC -O $URL
else
        echo "No downloading utility available"
fi
```

There are three lines of variable definition. $URL is a straight assignment of the URL, but the next two, $fW and $fC are assigned by $(command) syntax, which is a command substitution. The command is "which wget", which actually looks for wget in the path and if found, returns the full path of the command, i.e. /usr/bin/wget. Since this command occurs with the command substitution wrapper, /usr/bin/wget gets substituted in place, and thus gets assigned to $fW. Similarly $fC has the value /usr/bin/curl. Suppose wget is not installed, then which wget will return an empty string, and $() will substitute an empty string in place and $fW variable will be contain nothing.

The second part of the code snippet is a conditional statement, using the syntax if expression... then command1... elif expression... then command2... else command3... fi. It is testing for the existence and executability (-x parameter) of the full path of the command, and depending on whether wget exists, if it does, the wget command is executed and then it exits. If however, wget does not exist then it tests for curl in the elif line (short for else if), and if that exists, the curl -O command gets executed. If curl also does not exist, then the third command is executed, which is to echo a message "No ..." to the output. So it takes many lines of code, just to make a program more portable and more foolproof.

Many bioinformatics tools, deposited with the SourceForge open source project, are accessible from this open platform for downloading. However, due to the exceptional load, your file may be downloaded from another site, which SourceForge will redirect you to. This redirection is to distribute the traffic load to a site that is nearer to you from a network topological point of view. For example, if you are in Taiwan, it will redirect you typically to an equivalent local repository, in this case, the Taiwan National Network Center. In addition to using `curl`, there is the `-O` parameter and the `-L` parameter. `-O` means to use the filename that is in the URL for naming your downloaded file. `-L` means to allow the redirection to take place.

CHAPTER 7

Using a Bioinformatics Cloud Computing Platform

7.1 Simple Introduction to the Cloud Computing Platform

Cloud computing in recent years has been a very hot topic in the information industry, just like as a "school of thought", especially after the emergence of Google and Amazon, and their vast deployment of computers in the cloud. As a result of strong adoption of the cloud platform by industry and academia along with international organizations, the whole structure is actually very large, even though it can be divided into a very fine level of structural granularity, particularly for the ordinary user purchasing at 10 cents an hour an instance of a cloud server. But for a biologist, trying to understand the details does not seem to make much sense.

Let us put it in the simplest terms: the idea of the cloud arises from the concept of being able to "remotely operate" someone else's service, platform or hardware in order to complete whatever necessary work you have to do.

If the definition of this concept is pushed to the limit, then you are already in with cloud computing. For example, when we, numbering millions, use Google's Gmail service at any one time, we are collectively and individually operating in a cloud, because millions of us use their online mail editor, and send email to each other. In fact, in the whole process occurs on Google's servers, wherever they may be in the world. And even for our unfinished emails, there is a temporary copy on their server somewhere in the world!

In Google's MapReduce architecture, the job is distributed to all available hosts with Map, and when each is completed, the merged result is called Reduce. When you upload an email attachment, you will never know which server or where in the world you have uploaded it to. Your computer terminal's browser is actually controlling Google's Gmail service at your command. For example, many people are on social networks, like Facebook, Instagram or LinkedIn, or the traditional BBS (Bulletin Board System) electronic bulletin board PTT in Taiwan, which can also be considered as a cloud service. PTT provides exchange of information, which can also be used to write a message. Although it only has a text-based interface, BBS also possesses some cloud features, nothing less.

Well, have you ever set up a website? I think most people have the experience of using a web hosting service provider for their web content; based on the supplier's specifications, whether it supports PHP, Ruby, ASP.net or Python and other suitable computer languages, we can write Web applications which we can upload. This is essentially the application which uses the computational hardware offered by others to provide a platform for yet others to use as a platform for delivering to their end-users various network computing resources and computer functionalities.

With regard to the several applications discussed above, the simplest to understand is the BBS bulletin board. This technique emerged in the 80s of the last century. It supported text only because the only network connectivity at that time was the modem-based "narrowband" network connection. I can say, "The concept of Cloud should be considered as being more than three decades old". Because the speed of the network connection increased considerably, we can do far more things today. In addition, the emergence of globally accepted services, such as Google and Facebook, have together created the media hype that cloaked cloud computing with an unusual aura.

7.2 Amazon Web Service

Everyone loves and recognizes Amazon as the best network specialist online store selling books and groceries. But little do they know that Amazon is also one of the biggest and earliest industry leader to sell cloud services. Now it is the world's largest cloud service provider, and nothing strange about that! Why?

Amazon initially conceived the idea to sell cloud services because of "idle hardware resources". Website operators may support especially large transnational e-commerce sites, therefore they require many computer hosts sharing web content storage, network equipment, etc., to service each website visitor such that there is a good level of service provided. Each user's experience must not be affected by overload of hardware resources and the web browsing experience cannot be too slow. If the experience is too slow causing the website visitor's mood to be disturbed, how can the visitor be expected to buy things on Amazon?

So in response to the huge volume of visitors, Amazon set up a system with a large number of computer hosts to meet whatever the peak demand was. But during the low periods, many computers are not used, so why not sell these idle computing resources? So Amazon started a cloud business.

You may be wondering, why would anyone want to buy these computing resources? why not just buy a computer to have by your side all the time? If you are thinking in terms of the enterprise, with the Amazon computing resources that can be rented instead of buying computers, perhaps in terms of a company's operations, it might just turn out more economical. Buying a computer means that the company will have to recognize the purchased computer as an asset, which will depreciate every year as capital expense. Why not switch it to an operational expense?

The purchased computer to the company also incurs cost of ownership, including maintenance costs, especially for large servers, and the rental of floor space just to house these servers. Just these two issues alone are pretty nerve-racking for business owners. Then there is the computer staff's salary, labour and health insurance, pensions, etc. Together as overhead, they may add up to a higher cost compared to buying the computer itself! In contrast, if the lease of Amazon computing resources is to pay rental fees only, with no other hidden costs, it might just be cheaper. Concerning machine maintenance, Amazon is responsible for that round the clock. For academic institutions and small research institutes, the use of Amazon's services is relatively convenient. No one has to worry about hardware failures and other technical issues.

The Amazon cloud services sold is not a machine in the server room which you can access *via* the Internet after you register. The key thing here is the deployment of virtualization technology. This is a computer hardware architecture which evolved after the development of multi-core processor technology. The main concept is a computer host being able to support several sets of computers, known as Virtual Machines. Rather than a physical entity, each virtual machine is actually a separate computer job, which you can assign a different number of processor cores, memory size, hard disk space, and most importantly, they do not interfere with each other.

In the past, a company may need three physical computers, one for web hosting, another as a back-end database server, and the final one as a file server. Now we just use a single host virtualization technology and immediately all the three magically appear available, with space-saving and reduced power consumption. This Amazon's application virtualization technology sold is a cutting edge technology. The customer gets the right to use the appropriate virtual machine of choice, as a service which can be rented according to different budget levels available.

Talking about costs, which is partly what everyone is concerned about, Amazon is a for-profit proposition, and so, is there no cheaper way than to buy a computer? How about using more high-end machines? When we objectively analyse the current official offers, we get whatever we pay for.

In the past, the Amazon cloud computing services (Elastic Compute Cloud, EC2), were divided into six levels of hardware specifications: Standard, Micro, High Memory, High CPU, Cluster Compute, and Cluster GPU, which in turn were subdivided into different levels of computational power. An old subscriber can still use these service plans. Amazon now only provides few instances, customers do not have many choices like before, but it seems Amazon upgrade most of the instances and lower the price of some instances. As seen from the latest price scheme on AWS pricing page, Amazon's product planning is very comprehensive, almost taking into consideration the different needs of every possible user. Whether you need higher computational speeds, or you need a machine which has a large memory, whatever resources you need are made available as an online accessible resource.

For all the comprehensive range of hard disk space facilities being offered, it is clear that Amazon has put in a lot of thought. Large machines with high memory, are only supported on 64-bit operating system. To rent memory or CPU on particularly large machines, one would anticipate that the amount of data must be large. So a little more hard disk directly assigned and higher network bandwidth packaged will go a long way. If only a small microcomputer device is needed, the hard drive space can be optional. Moreover, Amazon has another service called Simple Storage Service, frequently referred to the S3 storage service: you need only buy more of this service as and when you need more disk space.

Having introduced you to the hardware specifications, we haven't even started to talk about the costs. Amazon services are based on the level of use of the machine room location, in terms of hours of usage. It is very much like Internet cafe billing methods, pay as much as you use. We can understand why different machines costs differently, but why would the locality of the machine room costs vary so much? It would not be much surprise if the variation of each region can be understood in terms of the differing costs of electricity, room venue rental, staff salaries, cost of bandwidth, etc.

Renters of Amazon cloud services can select the optimal cloud service location, typically based in or close to their own areas where they are located, or when the network is topologically closest to where the main users of their service are? So if you are a Japanese company, you may choose an Amazon site in Japan; if the users of the service are primarily in the Asia Pacific from the network point of view, a site location in Singapore might be more desirable. Currently, Amazon is based in the Americas, Europe and Asia, with a total of seven locations, namely Virginia, Oregon, Northern California, Sao Paulo, Brazil, Ireland, Singapore and Tokyo. The lowest hourly rate is Virginia and Oregon, the most expensive in Sao Paulo, Brazil. The lowest fees, which is in Virginia, is shown in the figure below.

In fact, these are just very simple cost estimates. Amazon has another charge: data transfer. And because the network bandwidth is very expensive, the charges can be pretty high. Say you want to have a fixed IP address for your server, this is

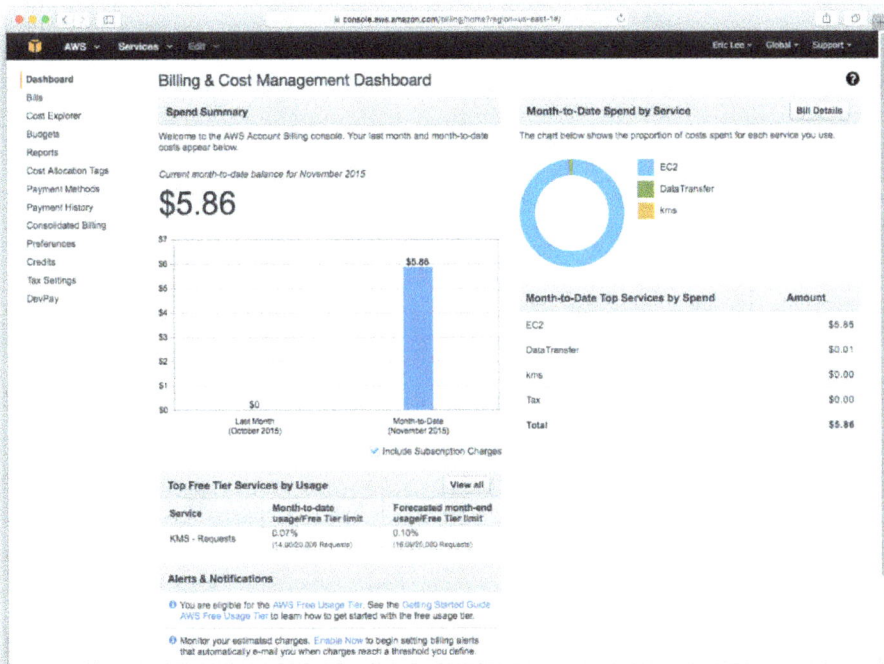

Figure 7.1   Just a reminder to the reader that there is no need to test Amazon's cloud services. When you start Amazon's services, you will be required to register your credit card number. Whenever a virtual machine is started, it starts billing your credit card and every download chalks up the metering of the bandwidth usage (except maybe at the microcomputer level, the machine may be temporarily free)

another fixed cost. But based on the hourly billing method, it is particularly suitable for some researchers who need a lot of computation suddenly, say to complete project. So why should he spend big money to buy servers when the need is transient? Actual users of Amazon's services do not need to worry about sunk costs. Even the installation of the operating system is expedited. Virtualization technology creates an operating system installation process that is so simple, complete with pre-prepared hard disk image files, directly loadable into the virtual machine, which you can use immediately and `ssh` into, shortly after the booting up, in a minute or so. You can also use images from other developers who have set up a good OS images, and immediately have a complete computing environment to operate in.

In the past five years, users abroad using Amazon cloud services have successfully developed a wide range of bioinformatics computing tools and services as described in scientific publications. Many bioinformatics tools were also directly developed on Amazon's cloud service platform. By contrast, within Taiwan or China, there is hardly any such research reports of cloud usage. In the case of Taiwan, perhaps, the broadband Internet access may not be fast enough, and bioinformatics data is simply too big, and transferring data online is to spend lots of unnecessary time and money. Moreover, users have alternatives in local resources, such as that in Hsinchu, Taiwan's national science city, where there is National Center for High-Performance Computing, with supercomputing resources and services easily available. Or perhaps, users are worried about data security issues, especially where human genome sequencing involves personal privacy issues and legal requirements for data protection.

7.3   Bioinformatics Cloud Computing Platforms

We have introduced the concept and definition of the cloud in the earlier section. The development of bioinformatics research tools shares many similar characteristics with the cloud. Many bioinformatics tools are readily available for others to use, to assist in their research discovery process. In response to the changing times, many tools already available, from shell scripting, Perl, or other programs are frequently combined and integrated into Internet-based Web-accessible sites. Whereas previously, only download sites needed to be supported, today, the maintenance of a website providing resources is a sophisticated process. To wrap many bioinformatics toolkits and software into a website service, the easiest and most popular way is to use with PHP, whose built-in syntax can directly execute Linux commands. With page templates, one can easily define a web form, for filling in data field parameters, as a very convenient graphical input interface. And together with information upload functionalities, these resources are as good as keeping up with the times of "cloud service".

In order to match the trend of Web 2.0, the latest and the best information can be shared, integrated community-based websites can be fully interactive, and standards in applications development are openly available. It is thus not surprising that a bioinformatics platform such as Galaxy can emerge. Galaxy was developed by the University of Pennsylvania as bio-computing architecture, so that the majority of bioinformatics tools can be operated through a browser interface. Previously, application service providers in bioinformatics have already gone through a massive boom and bust cycle, such as HeliXense, DoubleHelix, eBioinformatics.com, Entigen, BioLateral, and a slew of many others in the USA, during the dot.com days and the early and mid-2000s. In that Applications Service Provider model, users do not need to learn a lot of Linux commands, or how to install software. As long as the understanding of biological processes is there, you can immediately use these services. Workflow integration companies, such as Inforsense (bought over by IDBS) and KOOPrime, and myExperiment.org and Taverna projects, Pegasus, etc., have already proven such a model of integrated pipeline of bioinformatics services can work well. Today a new generation of bioinformatics service providers driven by the needs of Next Generation Sequencing (NGS) and genomics have emerged, including Eagle Genomics and those associated with personalized genomics services.

The most important point here is, as explained in earlier chapters, the need to establish a sound research workflow. This can now be seamlessly achieved on the Galaxy platform. Galaxy Workflows, using the concept of the process concatenation, allow for each tool, either the input data or output results can be pre-defined by the user, such that the output of one step of the process can easily become the input of the next step. No complicated setup process code is needed. Every tool of each step can be managed as a drag-and-drop object linked-and-connected by dragging to organize an entire scientific research workflow pipeline.

Galaxy is by far the most successful and easy-to-use open bio-computing cloud platform, not only convenient for the user, but also for ease of deployment by the system administrator on a heterogeneous Linux, Mac platform environments. (Windows versions used to be but are no longer supported.) Galaxy also has extended their functionality over the cloud, such that you can easily deploy to Amazon's cloud services. For system administrators who wish to deploy Galaxy on the Amazon cloud, there are conveniently instantiable OS image files (Amazon Machine Image, AMIs) which can be directly imported and adapted.

Moreover, tool developers who wish to integrate homemade applications can also do so very easily on the Galaxy platform. Galaxy has complete documentation with official tutorials explaining how you can build custom tools with XML file

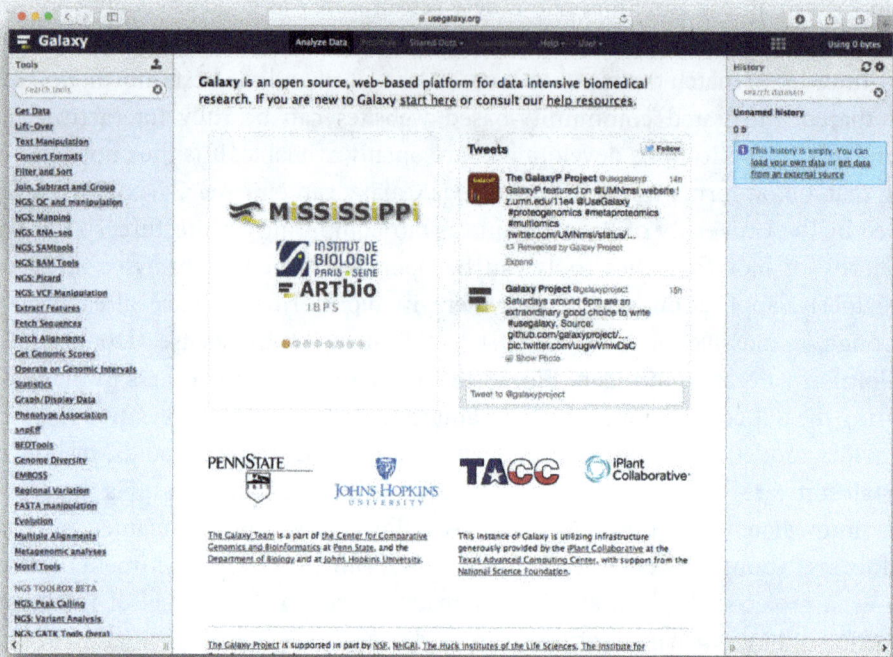

Figure 7.2 *The Galaxy master screen is very simple. On the left is the list of tools. In the middle is the work area and on the right is the schedule of work items. At the top are pull down menus for settings, preferences and sharing*

records (with the description of the tool's parameters specified therein), with utility tools for uploading into specified location and so on, such that simply restarting Galaxy will activate these tools for integration with other workflow tools.

In this section, we will explain the basic operation of Galaxy. After practising this interface to his heart's content, the user can decide whether he wants to build a personal private cloud or one for his lab, with a dedicated Galaxy server running all their applications that they need. The next section will explain the installation and important settings.

Logging in to use Galaxy services

Currently, Galaxy official has established two public servers, namely Main (https://usegalaxy.org/) and Test (https://test.galaxyproject.org/). The difference

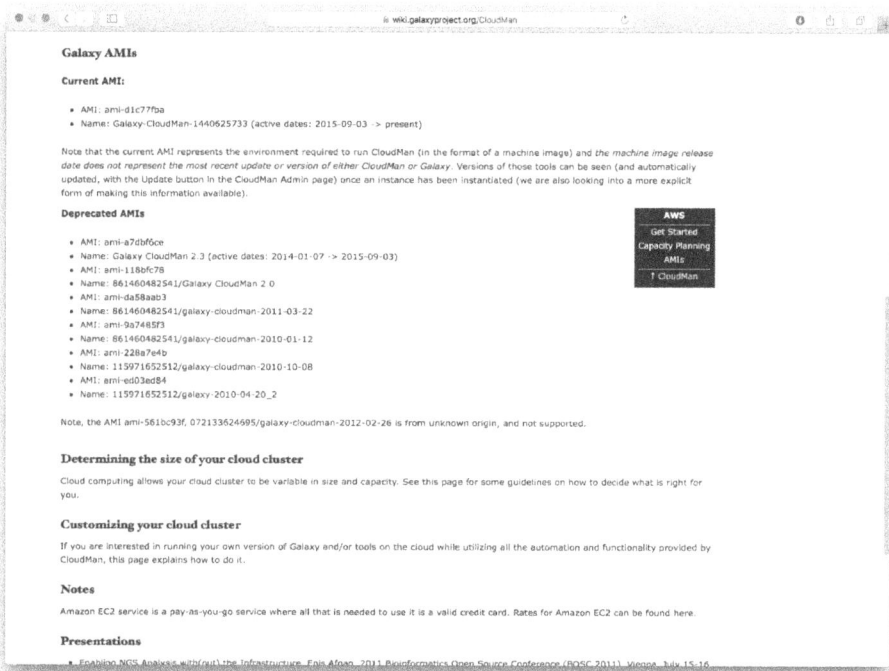

Figure 7.3 *Upon configuring settings in your Amazon cloud services, you can directly load the workflow environment with a Galaxy disk image. Although the Galaxy Wiki claims that there are three officially maintained AMIs, at the time of writing this book, I could not find them on the AWS*

between the two is that one is recommended for operational use, and the other is for testing purposes. Any applications in the developmental and testing beta stages, and not formally launched, will be placed inside the test environment. Usage of Galaxy does not require registration, and users can directly upload data to begin operations immediately. However, the biggest advantage is that after registration, the usable disk space becomes considerably larger.

Speaking about disk space, the master and test servers have different disk space usage restrictions. And the master allows up to eight jobs, which takes into consideration the overall system's stability. As Galaxy public servers are shared by many other users, especially when the user load is high, jobs sent for computation by the server are typically queued.

Table 7.1 *Differences between user accounts on the Galaxy master and test servers*

	Test server		Main server	
Type of account	Registered	Unregistered	Registered	Unregistered
Maximum Storage Space	250 GB	5 GB	250 GB	5 GB
Maximum Job	1	0	< 6	1
FTP Upload	Supported	Not supported	Supported	Not supported

Galaxy Registration is simple. It is not necessary to fill much personal informa-tion, just email address and password setting should be sufficient. For ease of use, it is recommended to register as a user.

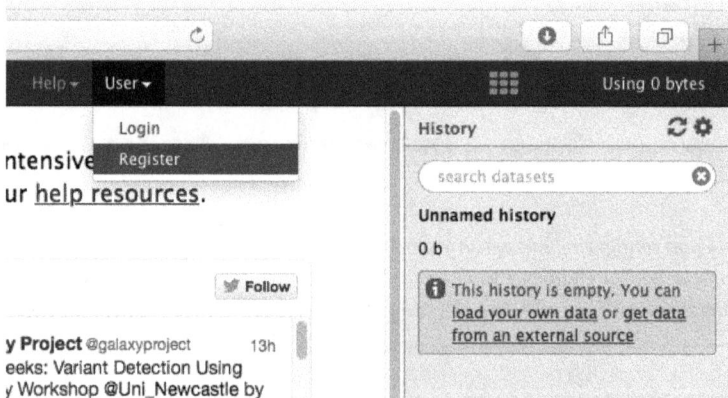

Figure 7.4 *For registration, click on the top menu — User, and select* Register

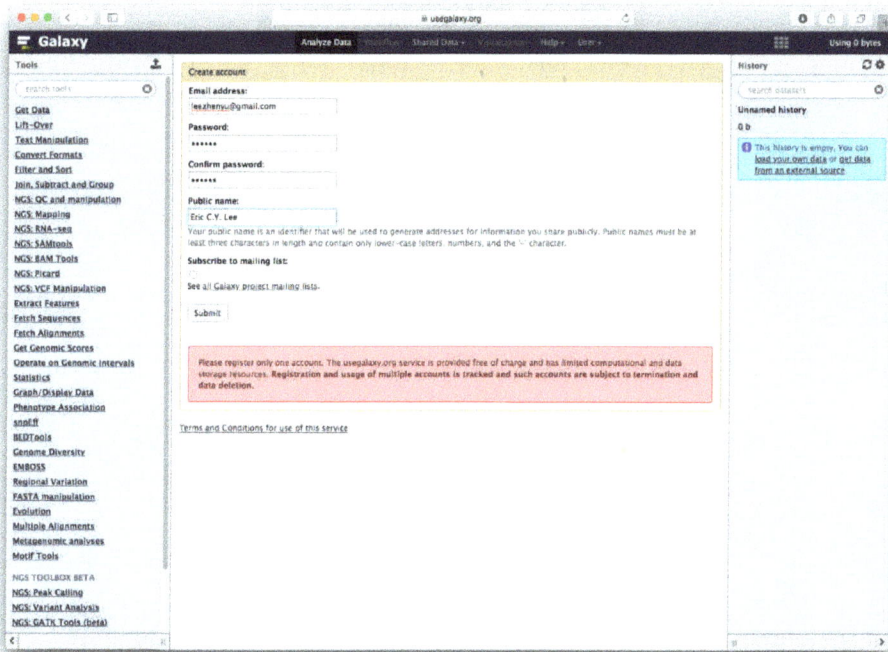

Figure 7.5 *Enter your email address, password and confirm the password, as well as a publicly displayed name. Decide whether to subscribe to Galaxy email notifications, and press the* Submit *button to complete registration*

Uploading sequence data

In bioinformatics research, the most important thing is your data. On the Galaxy site, only reference sequence data, which everyone uses, are preloaded. So the first step in the usage of Galaxy is to upload your research sequence data. The Galaxy site itself supports browser upload of data. This makes it very easy to upload small files quickly. Alternatively, you can paste the URL link directly (both HTTP or FTP are supported, provided that the site is not password-protected), and the Galaxy platform will directly download the desired files from the data server.

Galaxy platform also very considerably supports auto decompression functions. It is suggested that you upload FASTQ, FASTA or SAM files after ZIP compression. Because the format is a plain text file, you can usually achieve good compression.

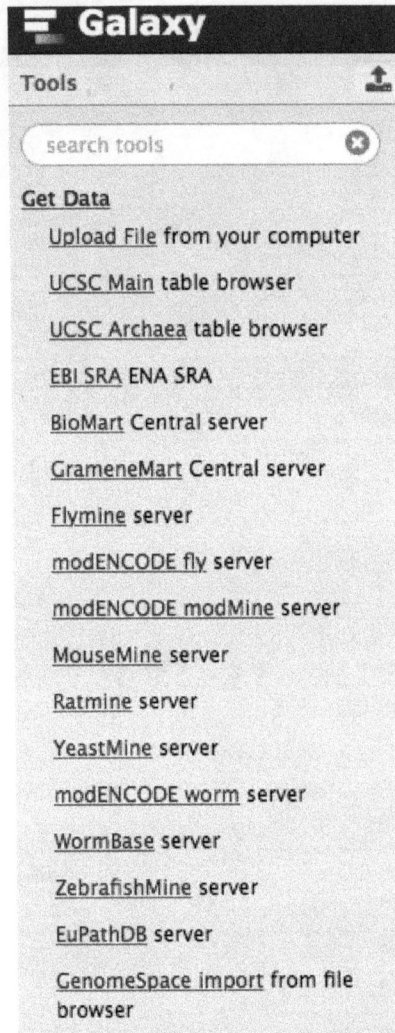

Figure 7.6 *On the left panel,* Get Data, *of the Galaxy interface is an* Upload File *option, which allows you to upload data from your computer. You can also specify standard reference and frequently used public datasets*

If the file to be uploaded is too large, it is recommended to use FTP for uploading instead, because your browser may crash if the data size is too large. In the Galaxy master account, for example, use the FTP protocol to connect to main.g2.bx.psu.edu. After Galaxy account registration, using a registered email address and password for the connection setup, you can typically use FTP to upload files. After the upload is complete, you can see a clickable link in the original upload page corresponding to the uploaded file, which you can select for usage.

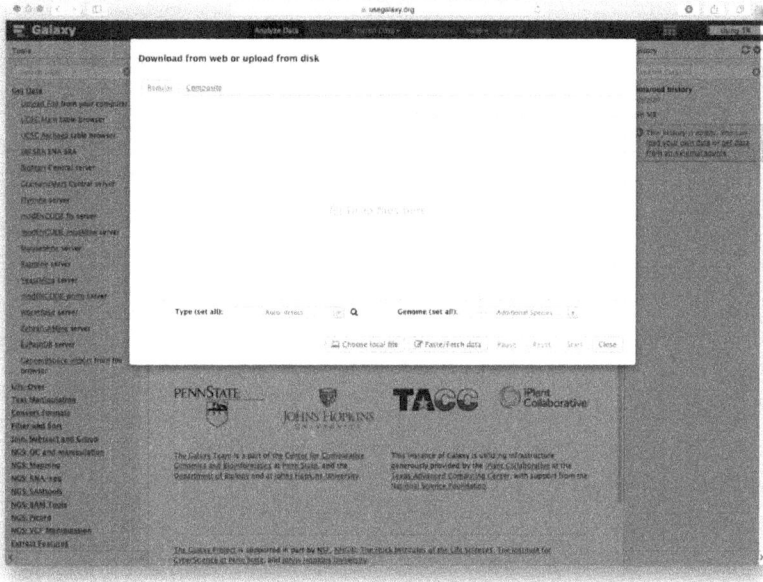

Figure 7.7   *In the* Upload File *dialog box, click* Browse Files *button, and the browser allows you to browse files in your folder in order to select the sequence file for upload*

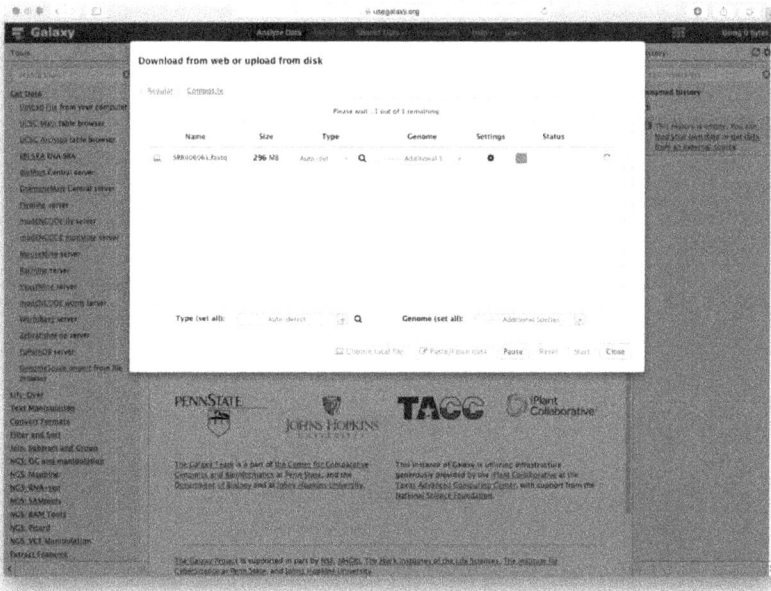

Figure 7.8   *Once you start uploading files, do not close or refresh the current browser window or else this will cause the upload to fail*

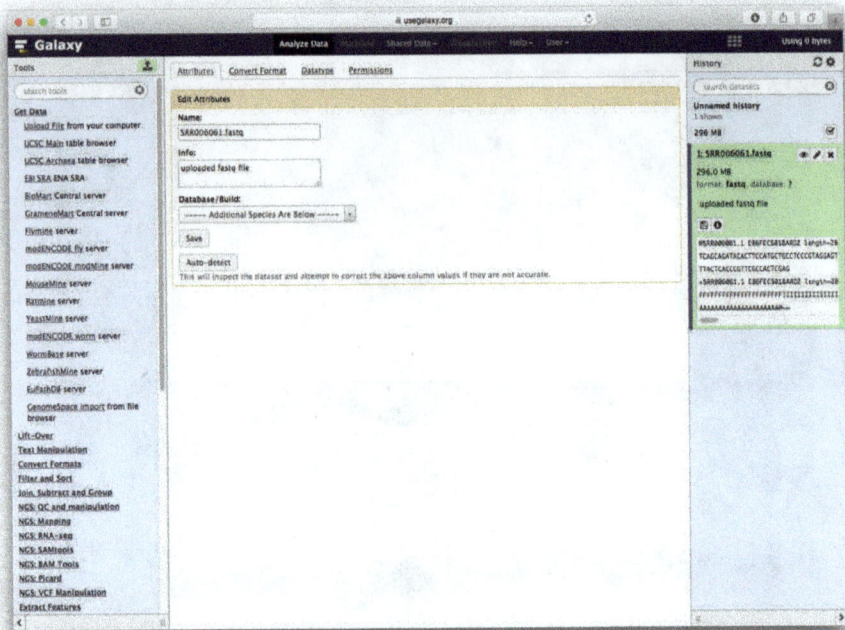

Figure 7.9 *After the upload is completed, on the right panel, the status will turn green, and upon clicking, the file size and other information can be seen. You can also edit the summary, or specify the sequence type (such as the human genome sequence). Henceforth, this set of data can also be downloaded back to your machine*

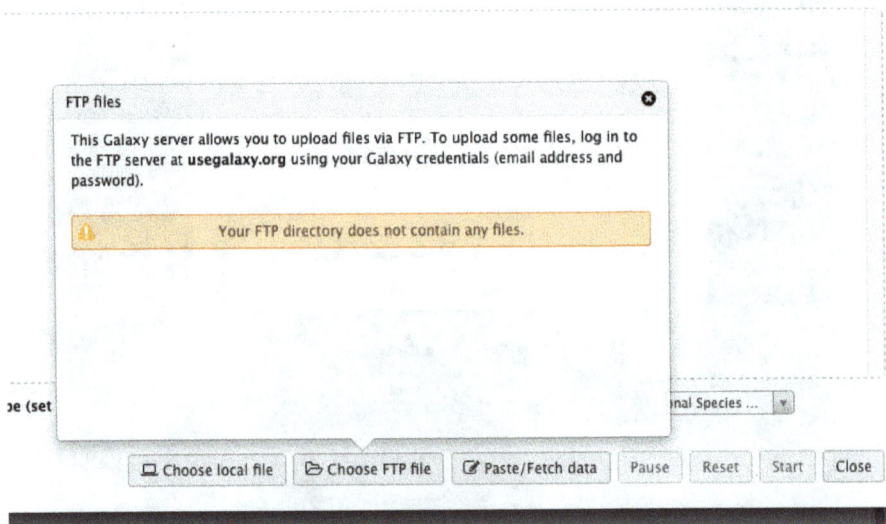

Figure 7.10 *In the middle of the Upload files window, there is a notification that you can use FTP to upload, but this is only after you log on to a registered account*

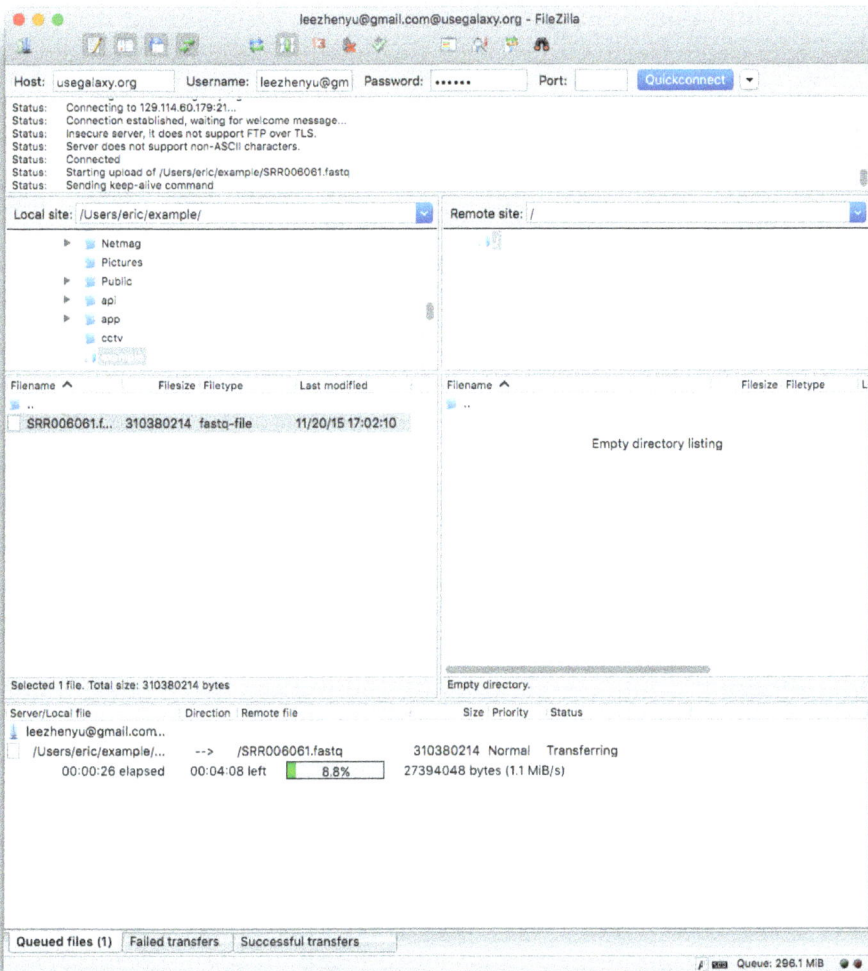

Figure 7.11   *To use FTP to upload, you can use a client, like the FileZilla program. To the main server, for example, the host location is main.g2.bx.psu.edu. Fill in the registration email account and password, and you can start uploading sequence data to Galaxy*

Regarding data acquisition, if you do not have any sequence data material but would like to test out and practise on the Galaxy platform, you can still access pre-analyzed or pre-uploaded sequence data, usually uploaded by official research projects, and import these data into your workspace. This is the fastest way to use Galaxy without uploading any data.

ame	Size	Type	Genome	Settings	Status

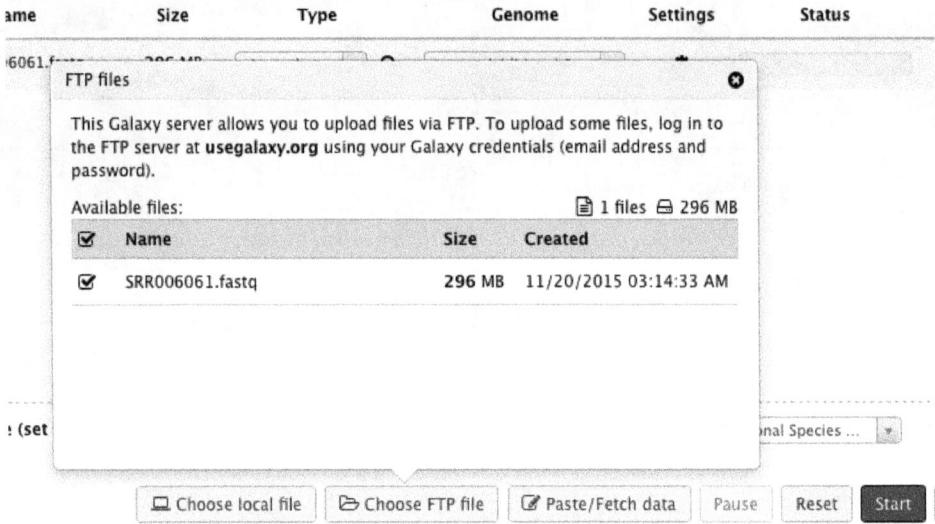

FTP files   ✖

This Galaxy server allows you to upload files via FTP. To upload some files, log in to the FTP server at **usegalaxy.org** using your Galaxy credentials (email address and password).

Available files: 🗎 1 files 🖴 296 MB

✔	Name	Size	Created
✔	SRR006061.fastq	296 MB	11/20/2015 03:14:33 AM

! (set   ınal Species ... ▾

⊑ Choose local file   ⮫ Choose FTP file   ✎ Paste/Fetch data   Pause   Reset   Start

Figure 7.12 *Once the FTP upload is completed, in the original* Get Data *page (remember to login), you can see the FTP-uploaded files. After pressing the* Execute *button, the file will be imported into the work area*

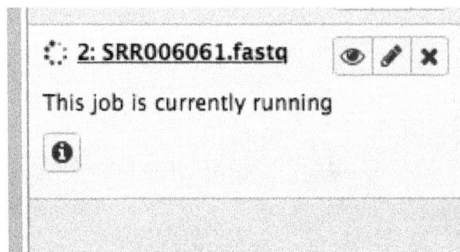

⟳ **2: SRR006061.fastq**   👁 ✎ ✖

This job is currently running

ⓘ

Figure 7.13 *Once importing is started for your FTP-uploaded files, the Galaxy main screen will be displayed in yellow and will indicate the job's progress*

In our examples shown here, our readers may not experience the same look and feel or be able to access the same data sets. That is because the platform manager may not necessarily pre-upload the same data files. In fact, even for the Galaxy main server, we can never be certain when a particular data set will be upgraded, changed, or withdrawn from support.

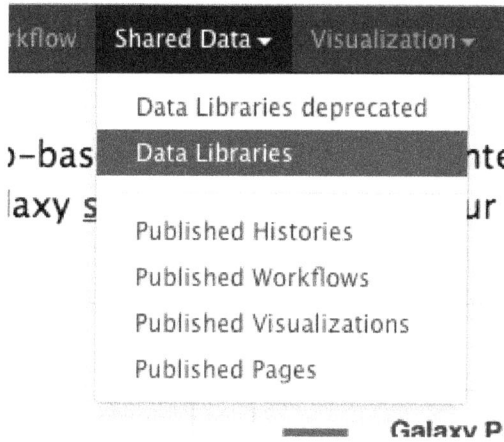

Figure 7.14 *From the top of the main screen of the Galaxy* Shared Data, *you can find within it the* Data Libraries *project, click to enter*

Figure 7.15 *The Data Libraries page is divided into two columns. On the left is the name of the dataset, and correspondingly, on the right, is a detailed description of the dataset. For demonstration, we chose Sample NGS Datasets, which are mainly three kinds of high-throughput sequencing platform output data from Illumina, SOLiD and 454*

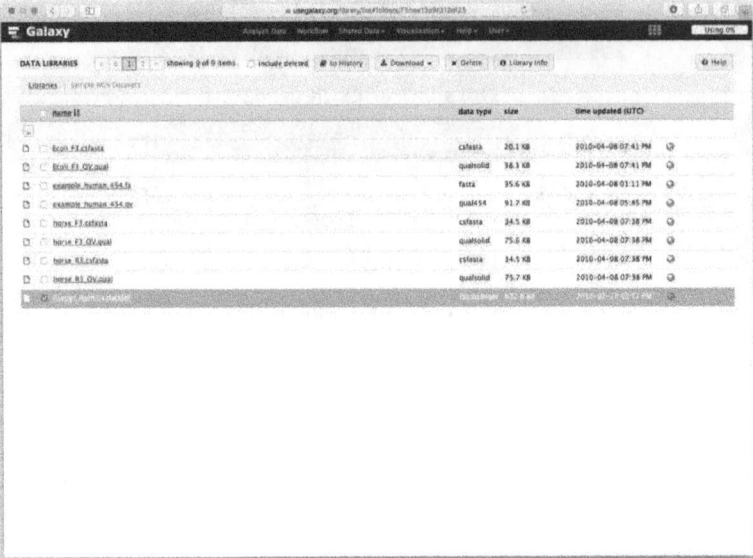

Figure 7.16 *Sample NGS Datasets come with sequence examples from a variety of platforms. Pick and choose the Illumina human genome sequence, check the corresponding box, before selecting "*`Import to current history`*" and press* Go *button to initiate the analysis*

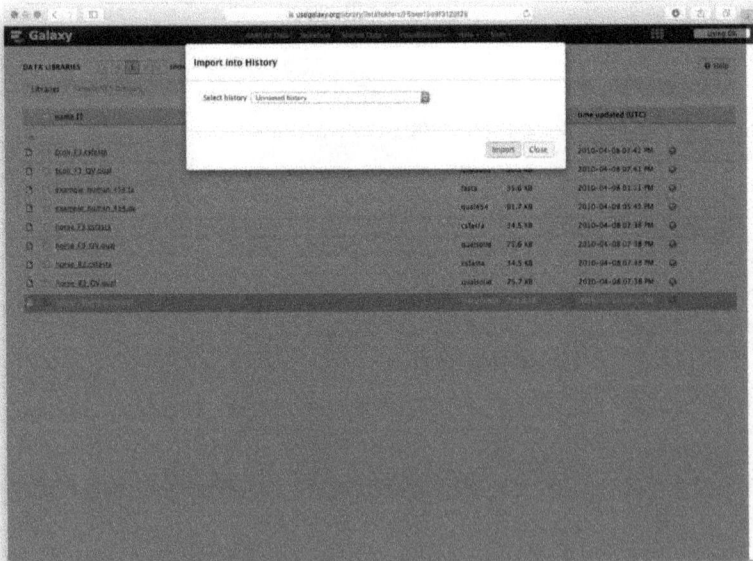

Figure 7.17 *Once the top of the screen shows the "*`Import into History`*" dialog in the workspace, you can click the "*`Import`*" button to start the dataset import process*

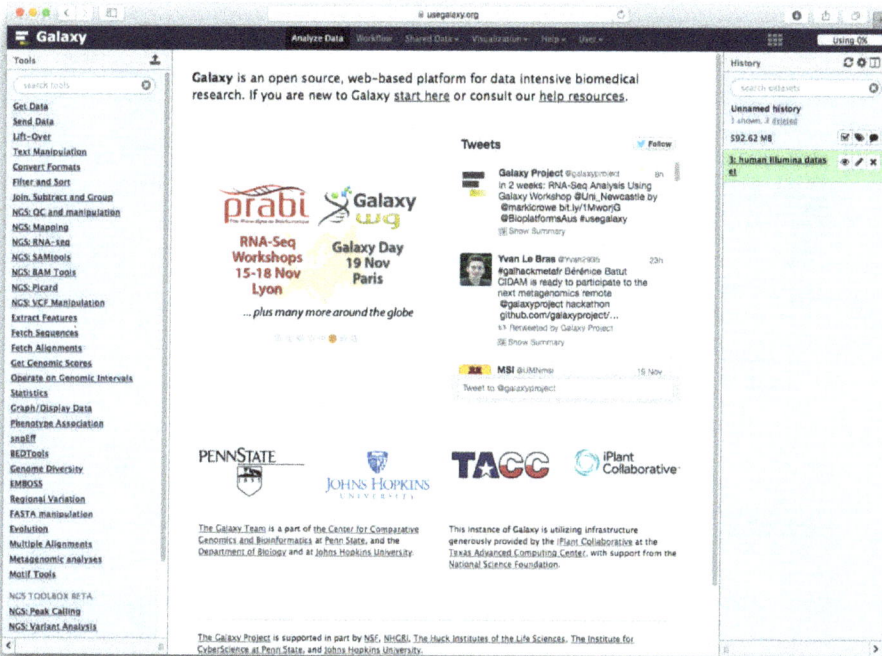

Figure 7.18   *Back to the main screen, you will observe on the right panel, just imported from the Data Libraries, the Illumina human genome sequence*

Sequence quality testing

Whatever originally executable on the command line, Galaxy can use a graphical interface to execute the same commands. And as everyone is almost always using a limited set of the same bioinformatics tools, the Galaxy development team has been able to keep up with ensuring all these commonly used functions are embedded in the official version of Galaxy.

As the first example of how Galaxy operates, we will use as demonstration what this book began with, that is to introduce a sequence quality analysis tool, FastQC on Galaxy. From the left of the main screen NGS Toolbox Beta, inside "NGS: QC and manipulation", at the very bottom, there is a link, Fastqc: Fastqc QC. Alternatively, instead of browsing, you can also use the search engine to find a quicker way to access FastQC program. As long as the selected sequence is detected, the execution will successfully generate an analysis report.

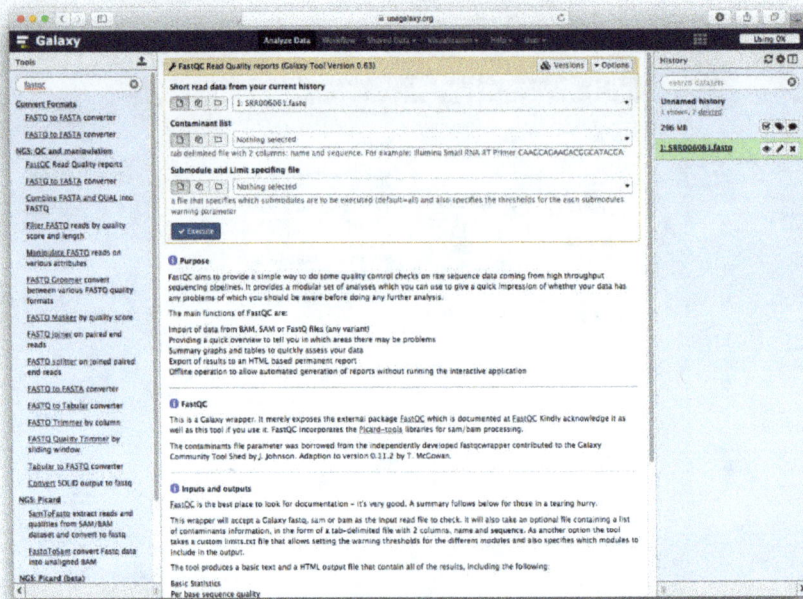

Figure 7.19 *Fastqc, on the Galaxy page, does not require too many parameters to be set. Just ensure that selected sequence is detected, and at most, include a convenient name to facilitate the title of the report, and then press* Execute

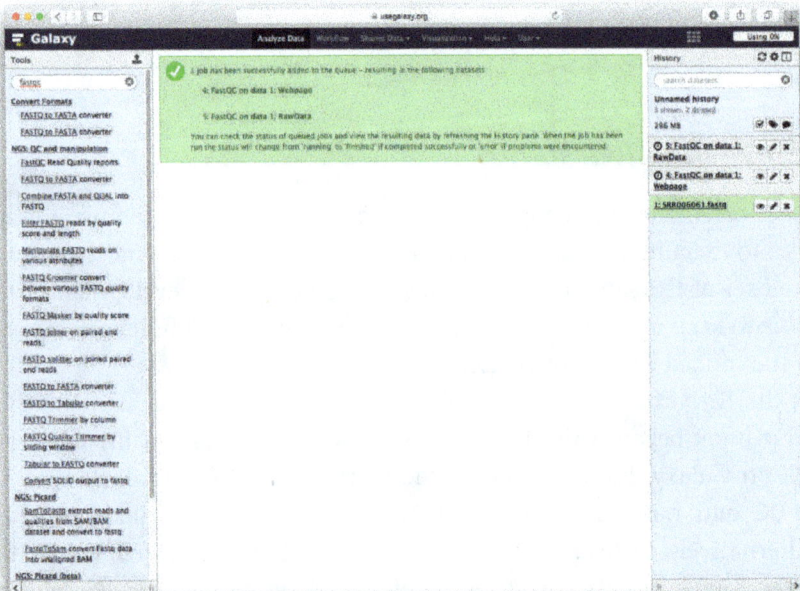

Figure 7.20 *Once ready to perform FastQC, you are directed to the right panel where the work zone is*

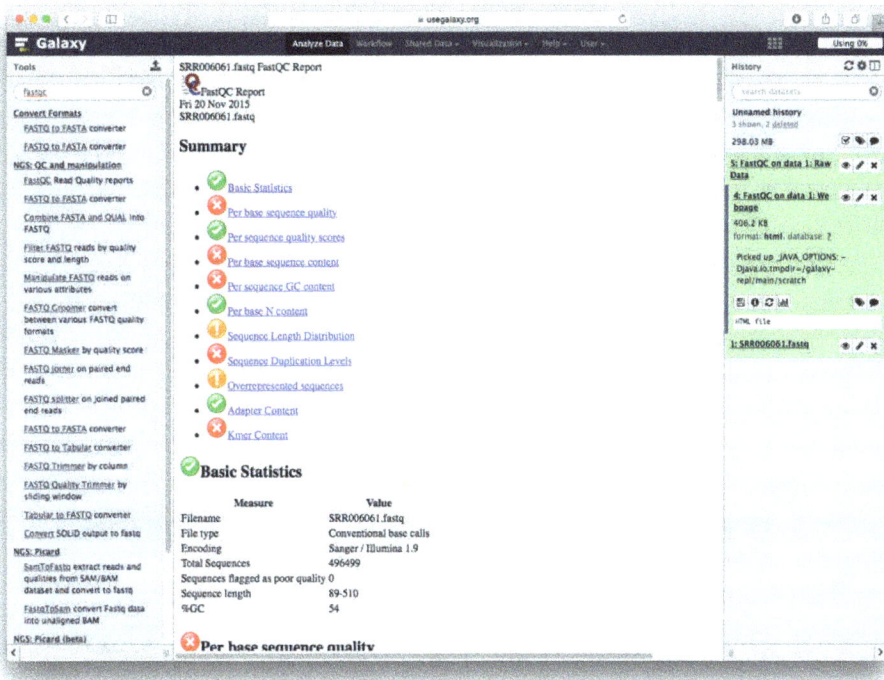

Figure 7.21   *The report generated by FastQC is no different from the stand-alone version of the results. The display of the HTML page itself is directly integrated and built into the Galaxy platform*

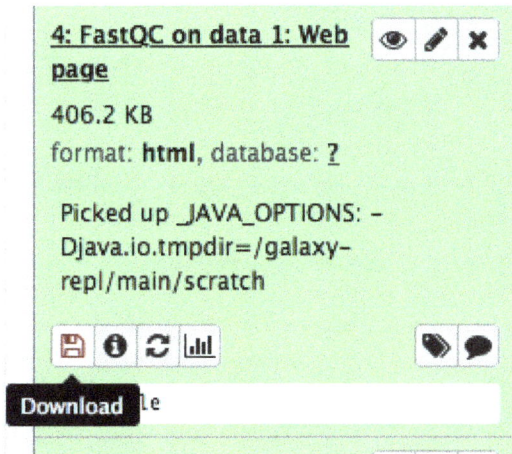

Figure 7.22   *As for packaged functionalities, Galaxy is also a very well-thought-out platform. Scrolling the browser down to the bottom, you can select options for directly downloading a ZIP format package of the results or an image*

In the sections of this book discussing sequence quality testing, I have introduced the FASTX-Toolkit for visualization purposes. Although FastQC with Galaxy has been able to meet our needs, we want to take this opportunity to make an educational point. While Galaxy is simple to use, there is still a certain logical order and data entry importation process, which needs to be followed closely, where even missing a small step will result in failure. So before visualization with a FASTX-Toolkit boxplot, you must first generate a statistical report. Such outputs of the FASTX-Toolkit plotting module is meaningful only if we accompany our analysis with numerical data as the fundamental basis. This principle is the same with each application within Galaxy.

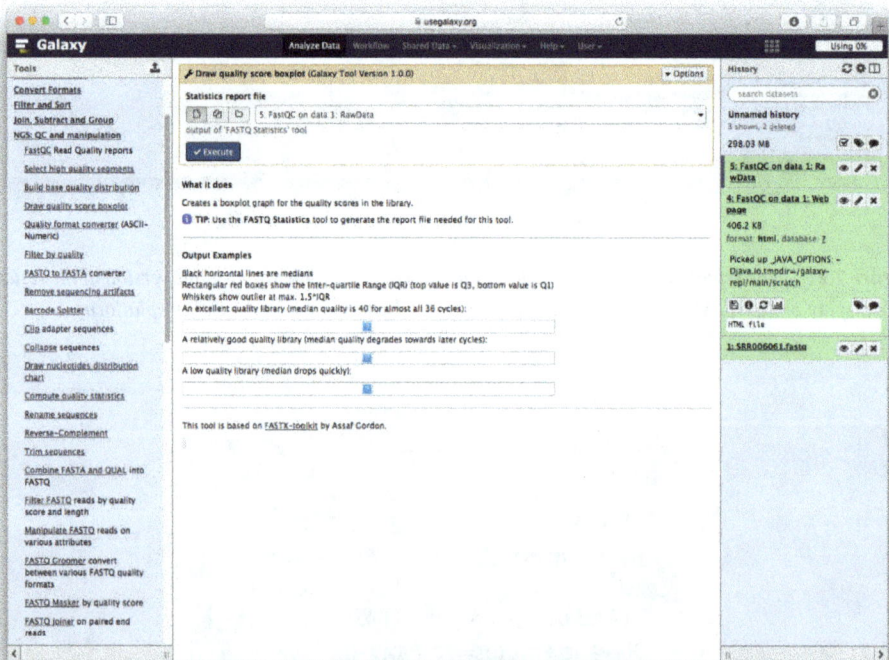

Figure 7.23 *Select the* Draw quality score boxplot *within the FASTX-Toolkit and a good selection of graphic formats for output of data (previously generated by FastQC). Once selected, you can immediately press the* Execute *button to proceed*

Figure 7.24 *Shortly after the job has been successfully submitted, on the right side of the main screen, an error message appears. Tapping into the tips and help messages, you are reminded to carry out* Compute Quality Statistics *first*

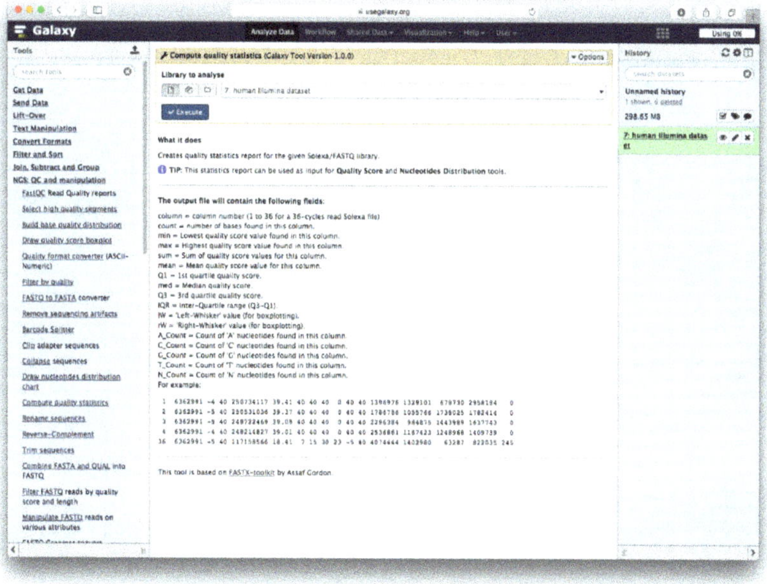

Figure 7.25 *On the left panel, under* NGS: QC and manipulation, *you can find under FASTX-Toolkit, the option for "*Compute quality statistics*". Select the correct sequence input and click* Execute *to generate statistics from a compliance report that was previously indicated to be required in the Plotting module*

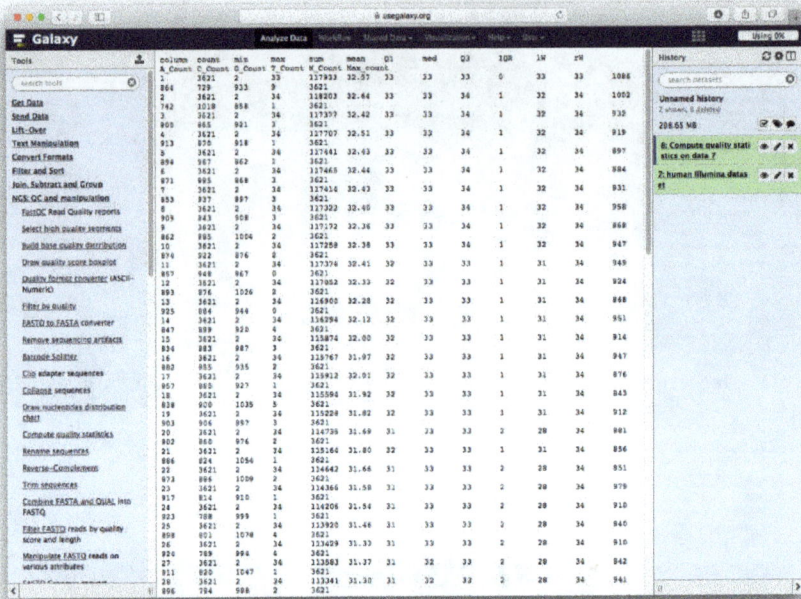

Figure 7.26 *The report of the step,* Compute quality statistics, *can be opened for inspection within Galaxy, which is identical to the Linux run-time command-line generated format*

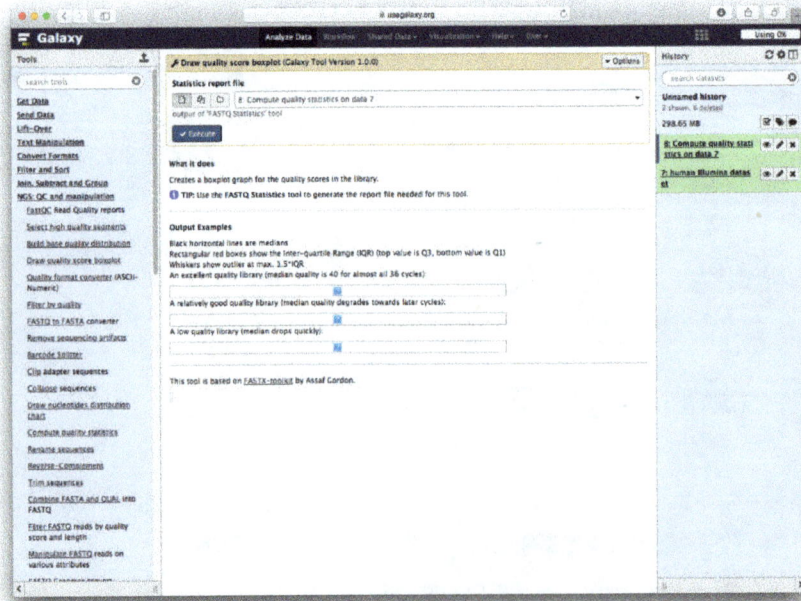

Figure 7.27 *Back to Draw quality score boxplot panel, just select the* Compute quality statistics *generated report, and press* Execute *again*

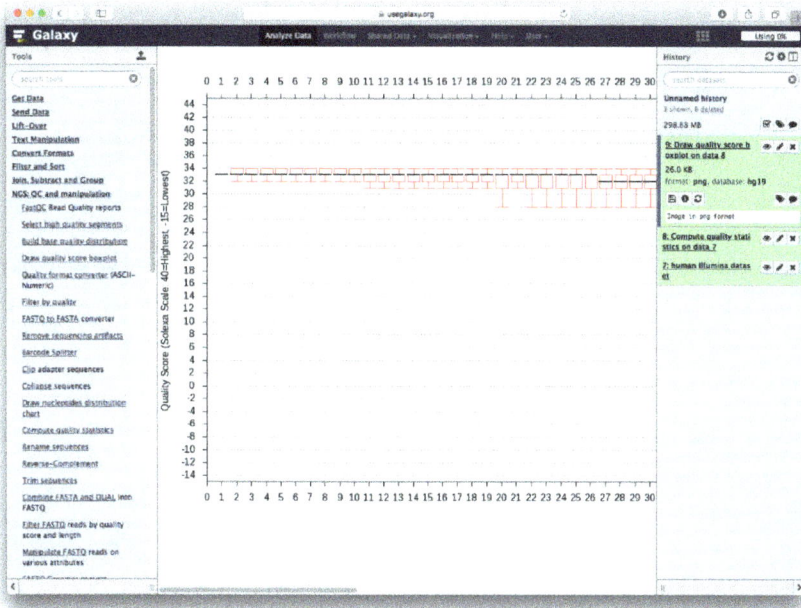

Figure 7.28   *Once plotting is completed, the output is as above. In order to facilitate a full map view, you can tap the two arrows at the bottom left corner and lower right corner of the screen to temporarily hide the list of tools and work history*

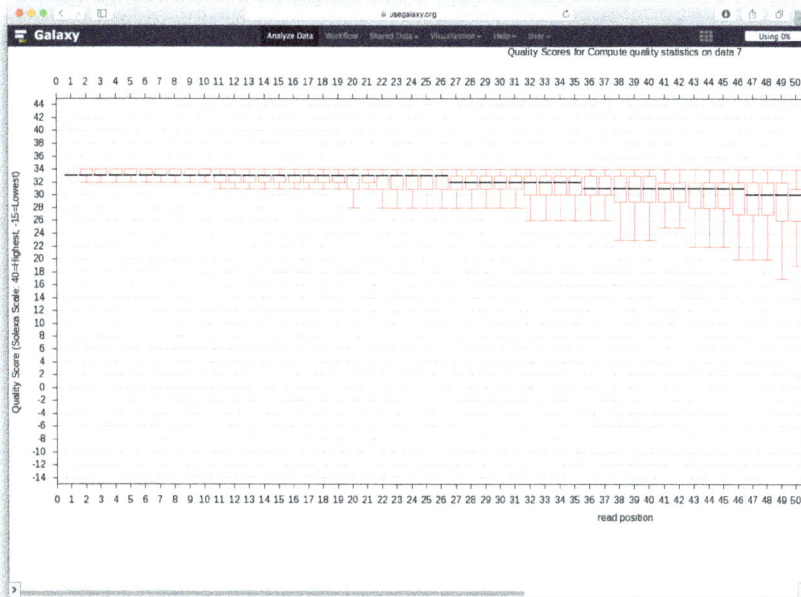

Figure 7.29   *You can scroll around to inspect the entire plot. Once confirmed that there are no errors, you can right-click to save the picture*

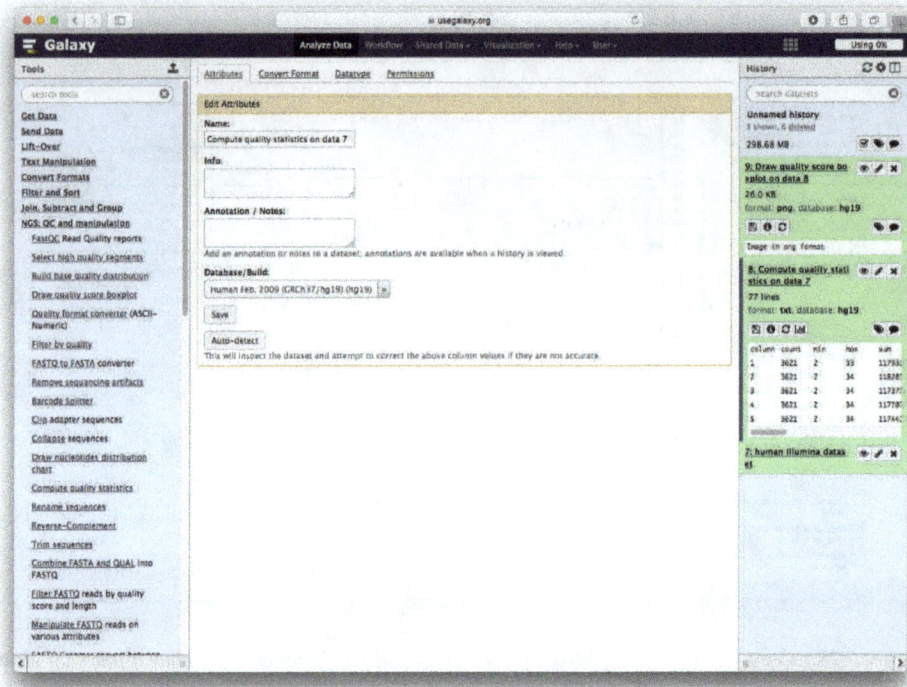

Figure 7.30 *You might ask how the chart title can be changed. The key is at the Compute quality statistics generated report files. Select the pencil icon from the results of the work process, modify attributes using "*`Edit Attributes`*" under the field "*`Name`*". Save and then repeat the Draw quality score boxplot drawing, and a chart title consistent with the name will be generated*

Execution of sequence alignment

Under the Linux operating system, to carry out sequence alignment using the command line interface is a big challenge for beginners. Especially for those who have never touched Linux before, ensuring that every parameter is completely and accurately keyed into the argument line is very difficult. Galaxy's greatest advantage is that all the parameters have been designed into a web form. Just fill in the correct values, and the correct commands with these arguments will be executed. So long as you know the concept of which tools to use for what purpose, it is difficult to go wrong. The official version of the Galaxy has integrated several sets of sequence alignment tools. Operating these applications is very much the same. Just ensure you select the correct kind of sequencing platform that was used in generating the sequence data, which you are currently analyzing. Examples found in this book mainly use Illumina's data. Note that Bowtie BWA is also available in Galaxy.

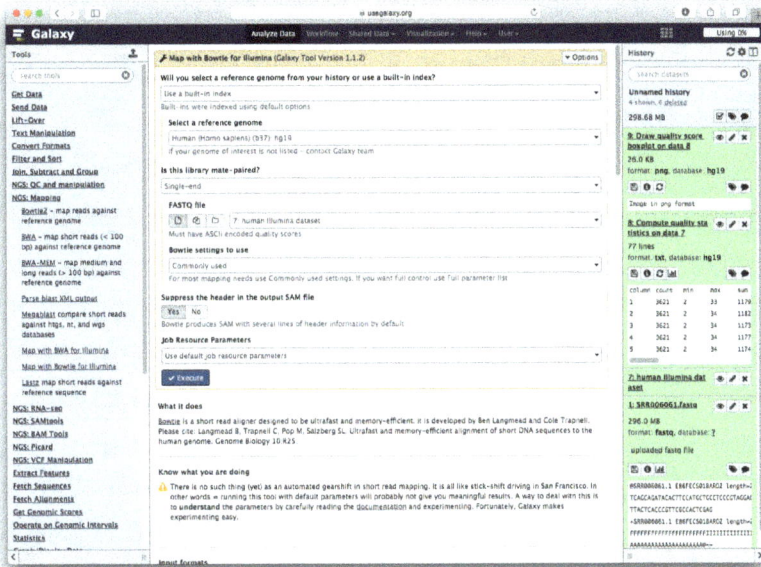

Figure 7.31 *In the NGS: Mapping tool for classification, under the Illumina category, there is "*`Map with Bowtie for Illumina`*". When clicked, the selected reference sequence can be compared with your FASTQ file without much detailed parameter settings. At most, check the output file for the SAM file, click a button, and press* `Execute` *to start the sequence comparison*

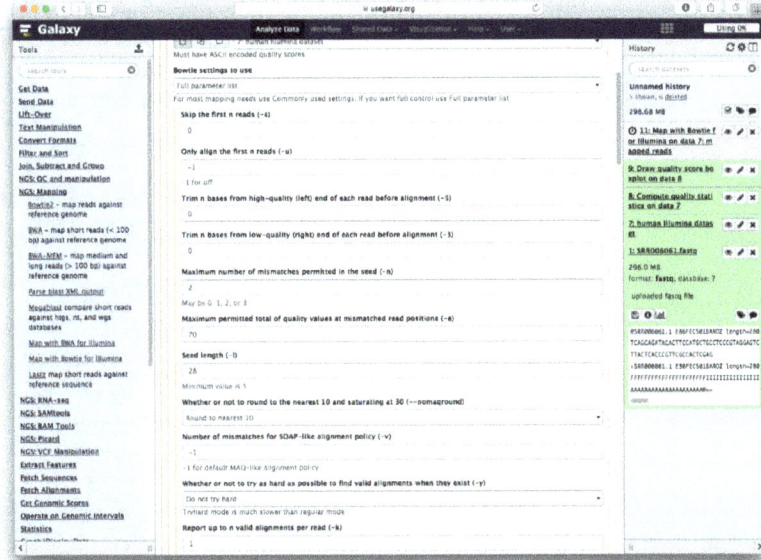

Figure 7.32 *If you need to adjust the detailed parameters, select the "*`Bowtie settings to use`*" from "*`Commonly Used`*" into "*`Full parameter list`*", all parameter fields will pop-up immediately, with full explanation. Note that the parameter controlling the number of processor cores cannot be adjusted here*

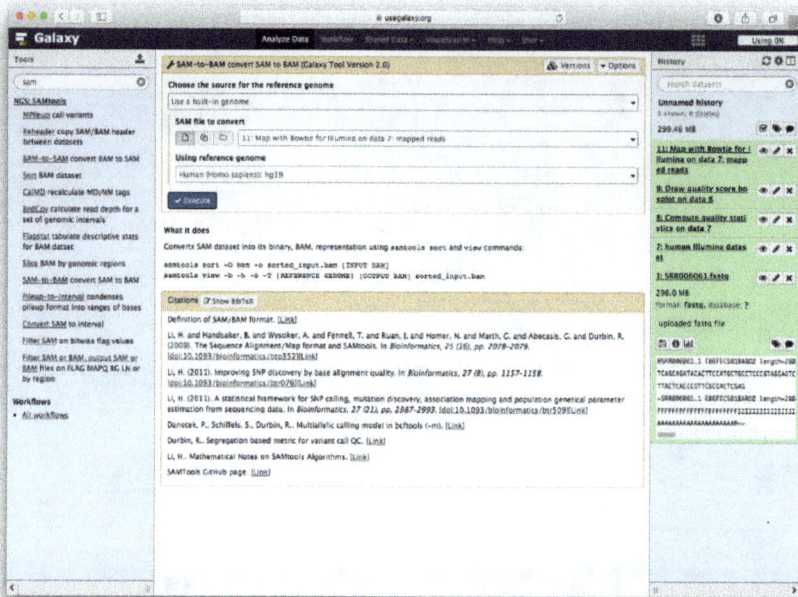

Figure 7.33 *In order to increase file read speed, switch the Bowtie SAM output file into BAM format. Select from the left tool area* SAM-TO-BAM, *then press* Execute *to start execution*

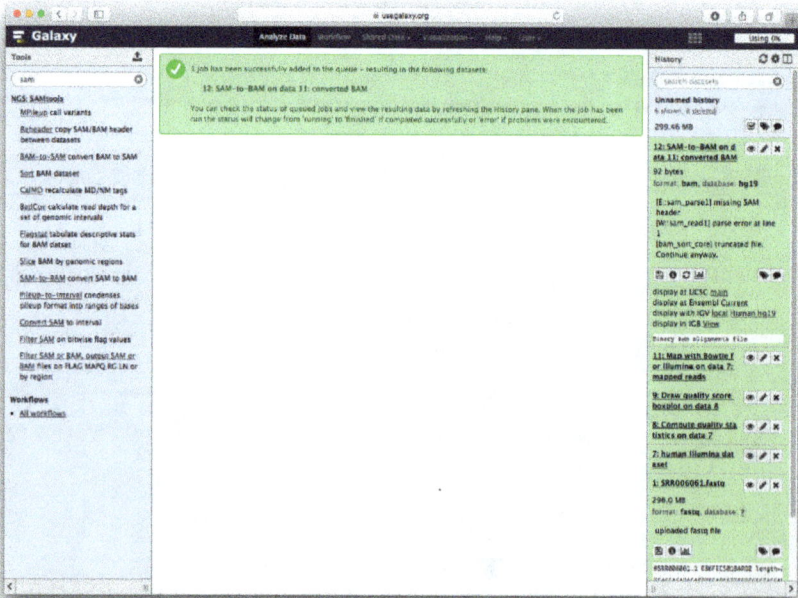

Figure 7.34 *Once the SAM file conversion to BAM format is completed, the file size is relatively smaller. At the bottom of the panel, there is the option* UCSC *or* Ensembl *to view the output. When clicked, it will load the BAM file into the Genome Browser*

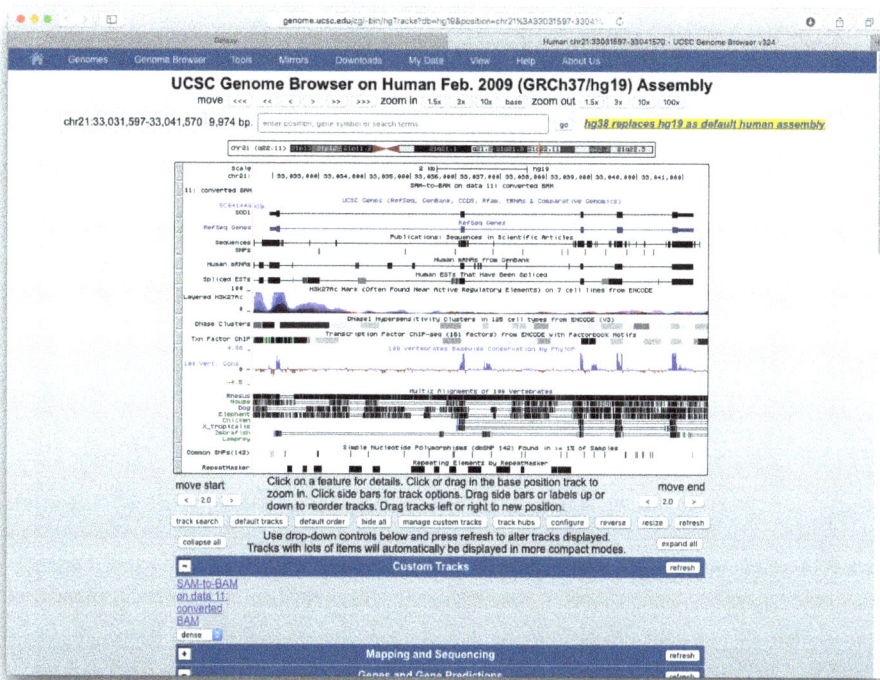

Figure 7.35 *Once the* UCSC Genome Browser *is selected, it will open a new browser window or tab and compare our results after loading the data on the UCSC Genome Browser*

Selecting other Galaxy servers

The official Galaxy server is typically highly utilized, and after a job is submitted, one must wait some time for the results. We can also try using other sites hosting public Galaxy servers. These sites may also host their own specialized integrated development tools themselves. Alternatively, there may be some official tools removed from support. For a list of other public Galaxy server, you can refer to the official Team Galaxy Wiki page (https://galaxyproject.org/public-galaxy-servers/).

When I was writing this book, it was typical to face a scheduled sequence alignment job which does not run even after a day. Perhaps changing to another country server, which is not so heavily loaded, the sequence alignment job may well start running immediately.

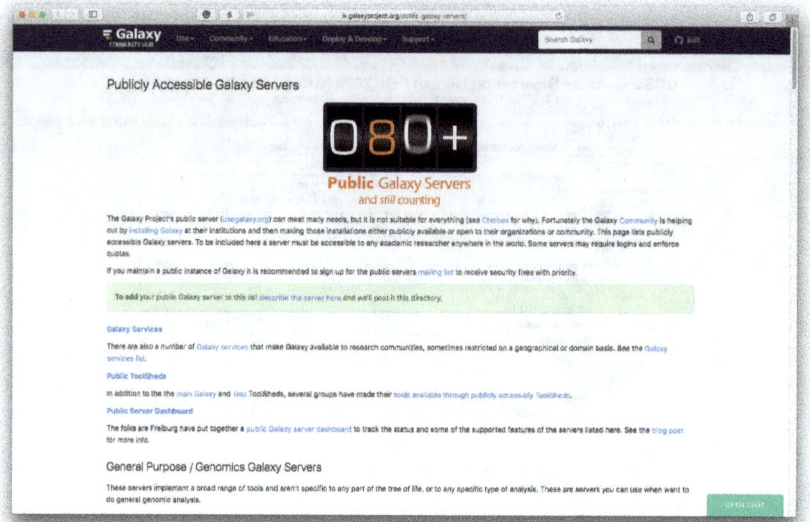

Figure 7.36 *Galaxy's Wiki page containing a list of other unofficial servers in the global initiative to provide public accessible Galaxy servers. For each server, there may be slightly different characteristics and use restrictions. For example, some servers force users to register before usage is allowed. Typically, there are simple introduction and help pages*

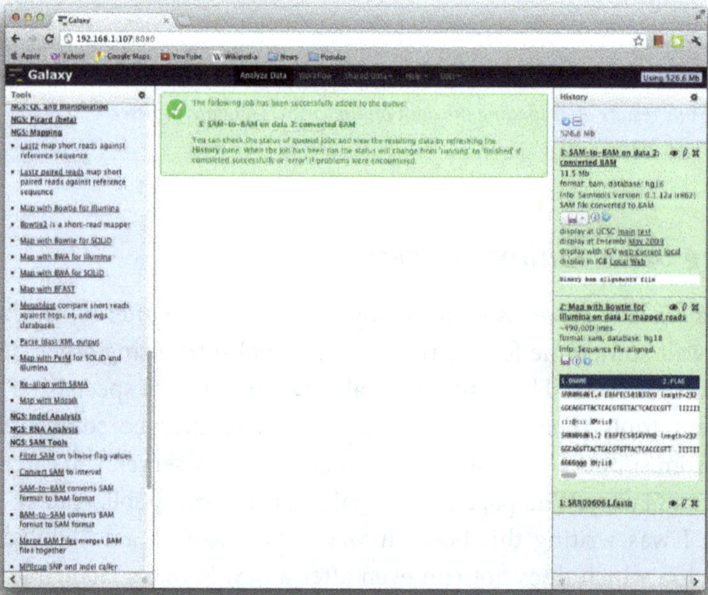

Figure 7.37 *When I set up my own Galaxy server after downloading the latest version of the Galaxy installation package, I did not expect that in addition to the previously mentioned UCSC Genome Browser and Ensembl, there are two other kinds of Genome Browsers, IGV and IGB. Maybe by the time this book is published, Galaxy's official master server will also begin to offer the other two kinds of Genome Browser*

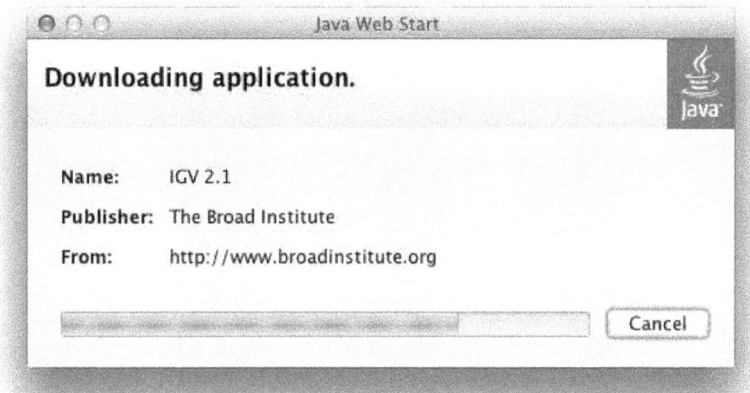

Figure 7.38 *Click on the link* IGV *and it will start the Java machine to call the main program of IGV. If not previously installed, it will automatically initiate download for installation*

Design and use of research workflows

If you think it is too difficult to write Shell scripts on your own, Galaxy's Workflow feature should meet your needs. Differing only in that the Workflow is so designed that it cannot be exported into Shell Script. From your own designed workflow, you can keep it private or make it publicly available. Importing someone else's workflow and modifying it to meet your needs is also not a problem.

Establishing new research workflows

Figure 7.39 *Tap on* Workflow *at the top of the main Galaxy screen to enter the Workflow page. There are currently no Workflows. Press button "*Create new workflow*" to create a new one*

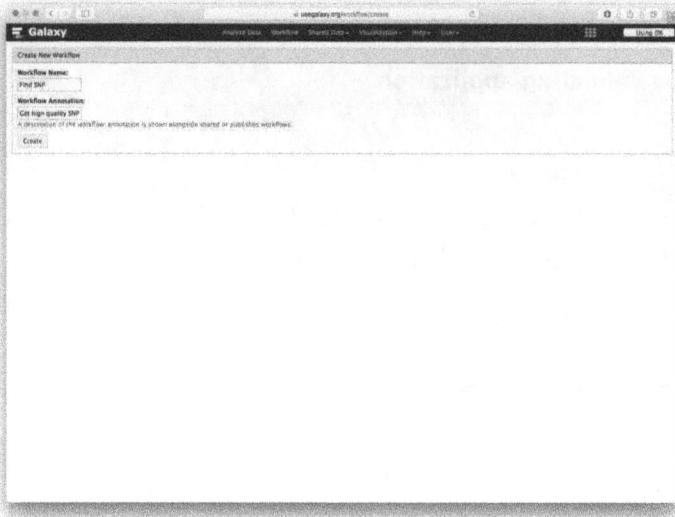

Figure 7.40 *Give this new Workflow a name and add some informative description. If you are going for public release, of course, the more detailed the description, the better*

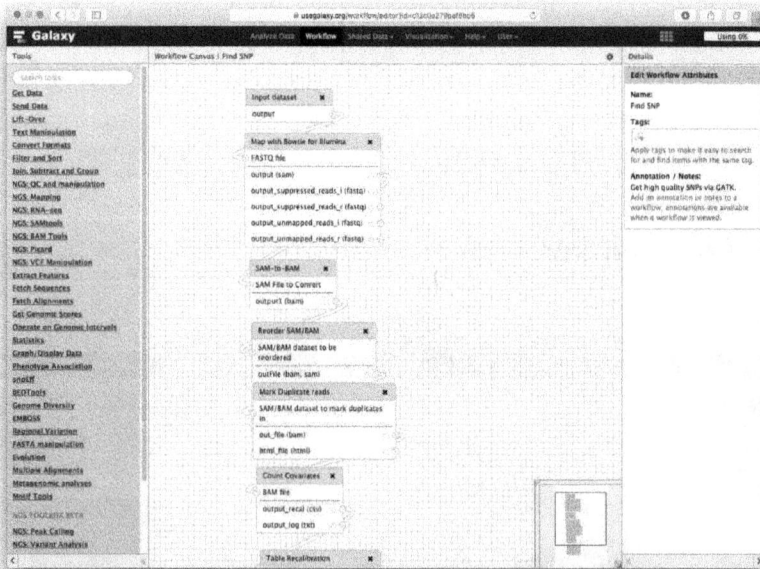

Figure 7.41 *On opening the editing screen, the central work area is a blank grid. On the left are the tools; on the right are every detail of a tool set. The adjustable parameters are similar to what you normally encounter in Galaxy. You can also write your own notes and save them in the Annotation field, equivalent to comments in a program. Within each tool, you can use your mouse to drag a way to specify the output data from one program to the next program in your workflow pipeline. Usually, it will also automatically prompt you for the output and input formats*

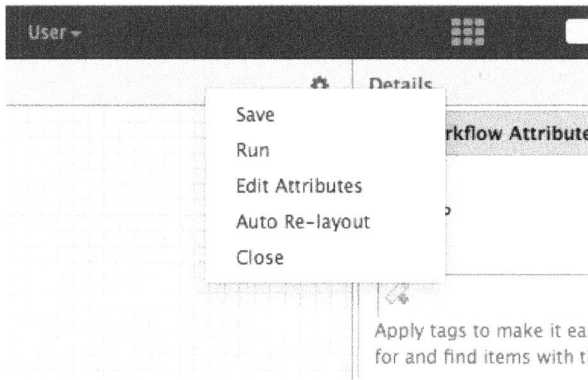

Figure 7.42 *Throughout the Editing process, you can press the* Options *button on the top right of the editing area, and accordingly, to* Save *the workflow,* Run *to initiate execution,* Edit Re-layout *to allow Galaxy to automatically help you put all the tools in a neat row, or* Close *to terminate the application. Let us select the workflow here for a demonstration*

Sharing and publishing process

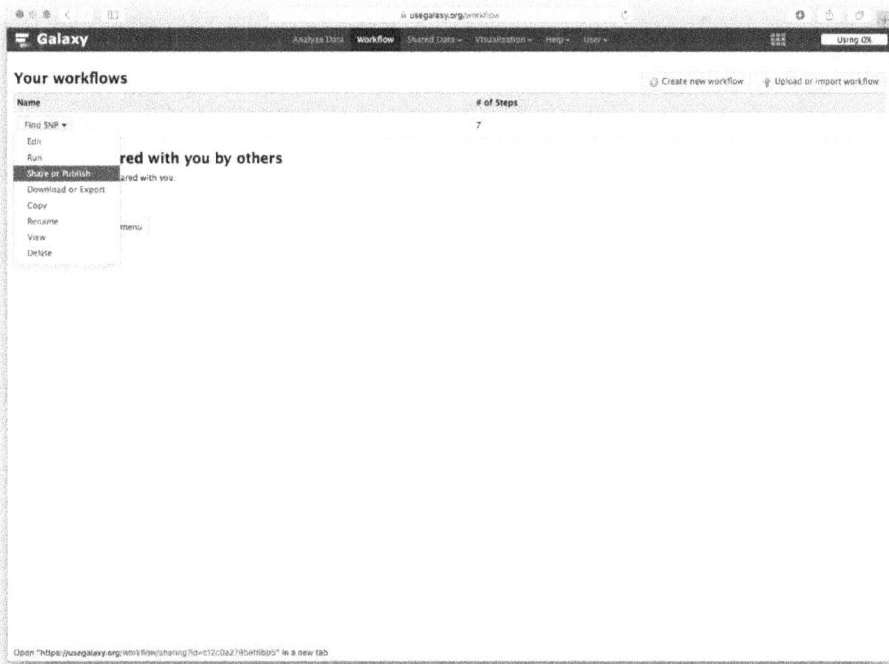

Figure 7.43 *Back to your workflows panel, the newly created Workflow will appear. Click on the Workflow name, tapping* Share *or* Publish *operations, will help you to start the sharing functionality*

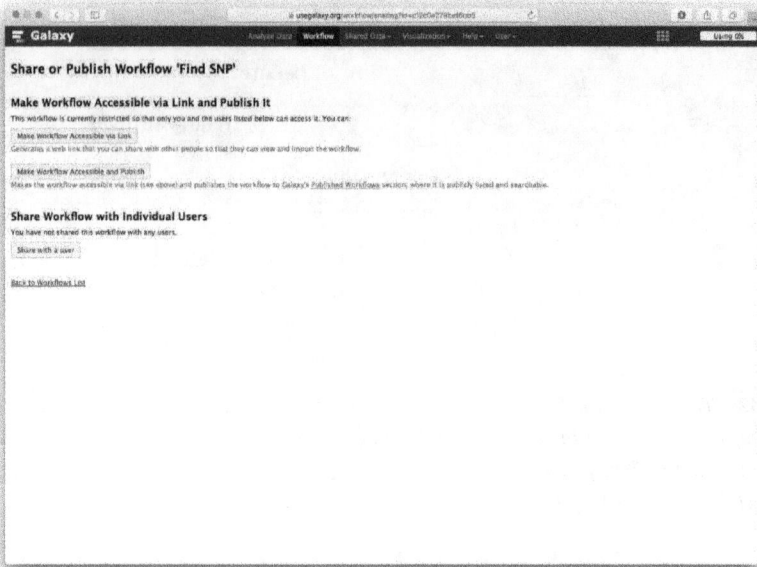

Figure 7.44 *Sharing Workflows allows several modes:* Make Workflow Accessible via Link, Make Workflow Accessible and Publish, *as well as to allow sharing to specific users (*Share with a user*). Select the desired sharing mode*

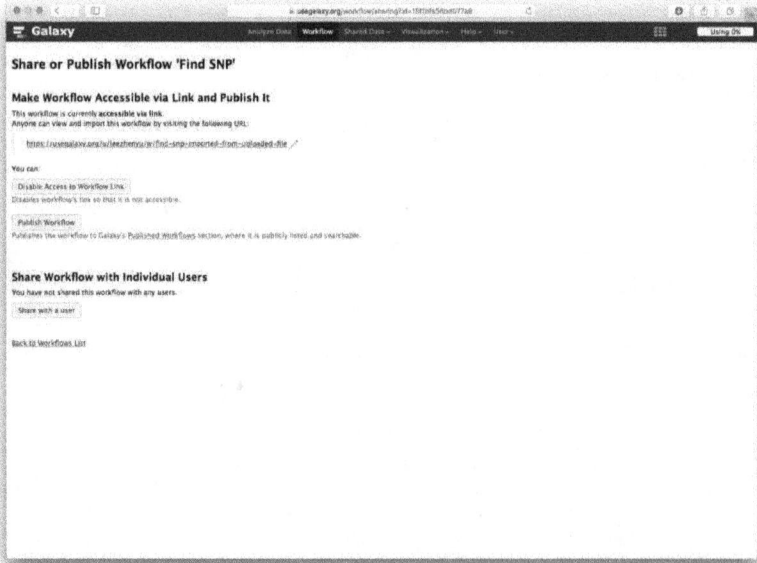

Figure 7.45 *If you have chosen to share the link, it will immediately generate access to your research process hyperlinks. Any person can see this link to your Workflow*

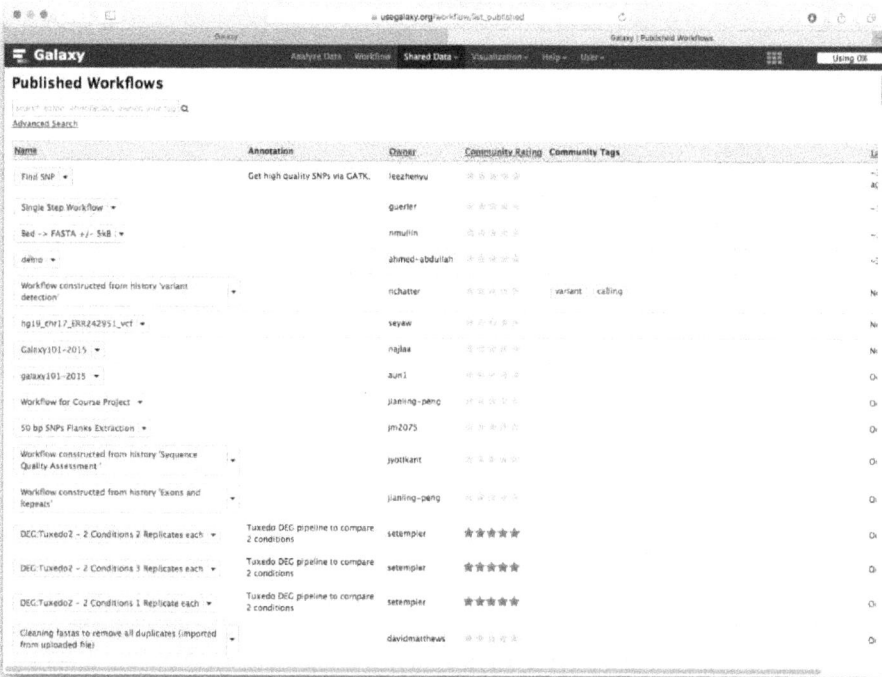

Figure 7.46 *Pressing the button to share published research workflows will immediately make public your workflow, and insert it to the Galaxy Published Workflows page. This action is completely without official audit, and directly appears on the list. Sorted by post time, the latest workflow in front. So after you press the button, usually the first one is your workflow. If you want to cancel the sharing, you can simply turn off sharing in your workflows page*

Figure 7.47 *To share the workflow to a platform-specific Galaxy user, just enter his email address. Please note that they must be in the same Galaxy server*

Execution of research workflows

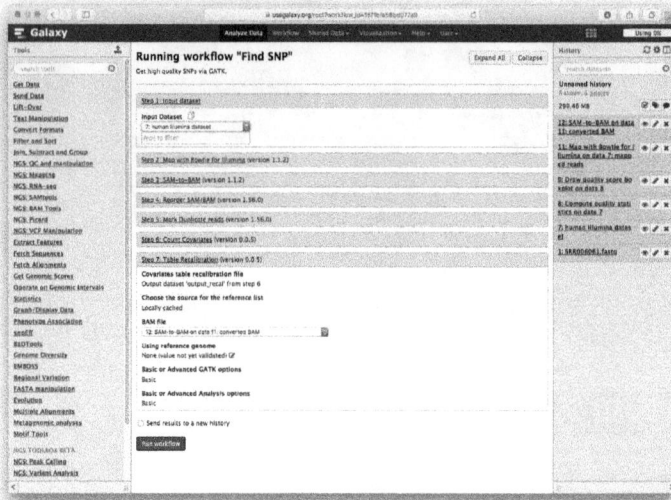

Figure 7.48   *From your Workflows screen, selecting* Workflows Run *will list all the processes. From the first step of loading the input data to the last step, there will be an* Execute *button. Pressing the button will help you through the entire workflow, as long as no particular process in the Workflow has triggered an error*

Downloading or exporting research workflows

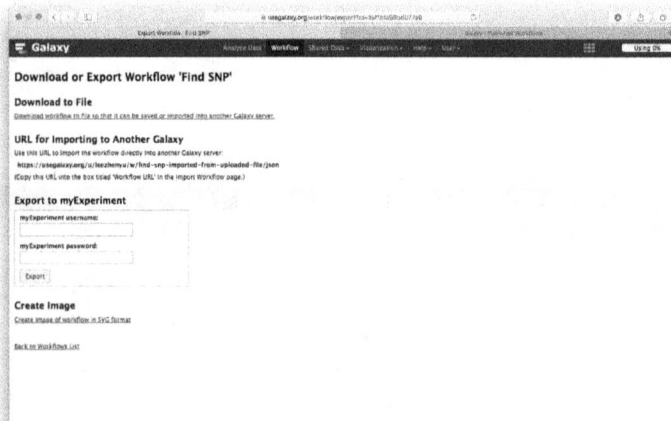

Figure 7.49   *For selecting* Download or Export, *there are three approaches for download or export research workflow. The first is a direct file download: downloaded files can be uploaded to other Galaxy servers and imported directly. The second way uses the special URL generated. On the other Galaxy platform, directly key in the URL to import, and it will grab the file from the server on your behalf. The last one is to import into a myExperiment website. To do this, enter the username and password of your myExperiment website account*

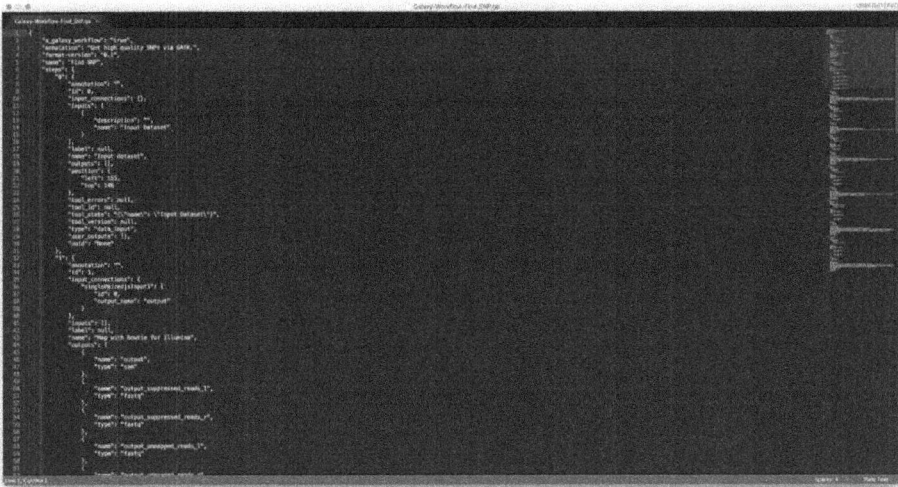

Figure 7.50 *After downloading the file, start a text editor, and the contents may look like JavaScript, but in fact, it is the JSON format. You can also define or edit the value of each parameter field as well*

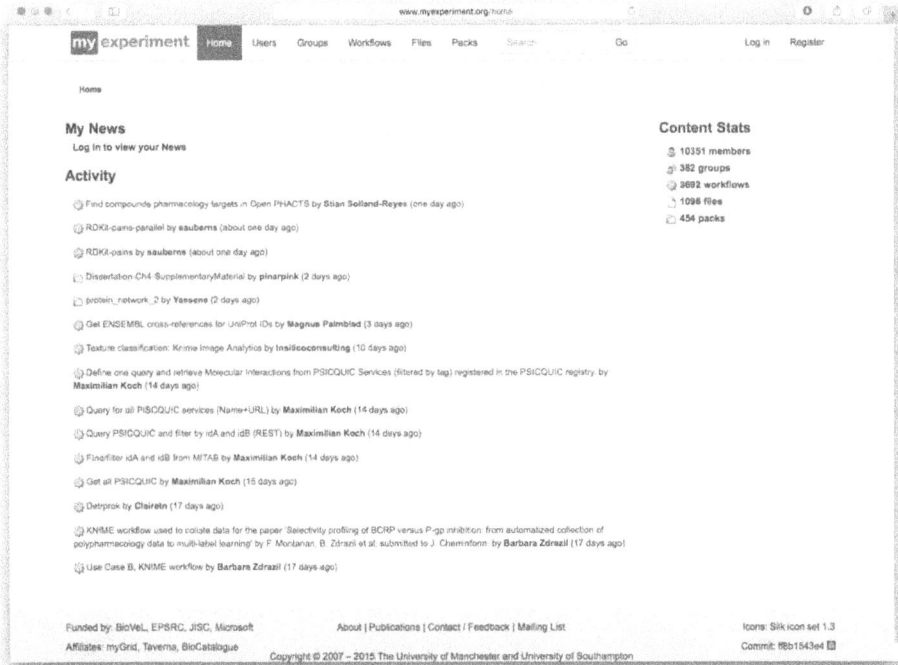

Figure 7.51 *myExperiment is a bioinformatics website for designing workflows (www.myexperiment. org). There is a more beautiful colour interface than the Galaxy interface, and you can also share or refer to the work of others*

Importing research workflows

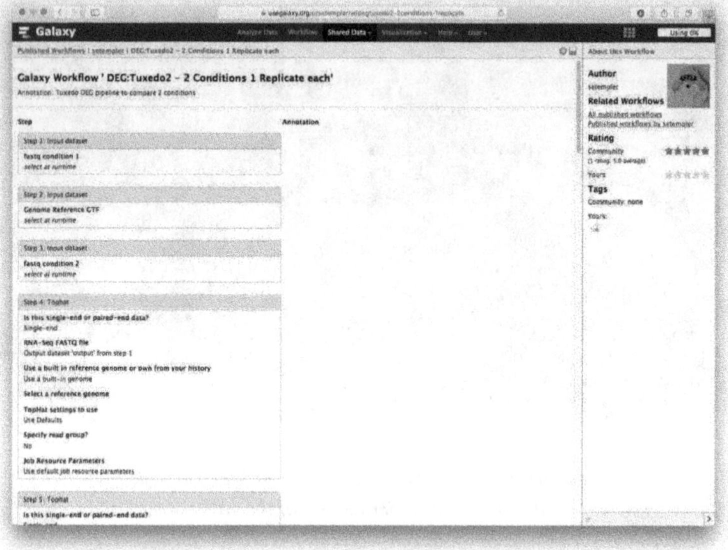

Figure 7.52 *From the* `Published Workflows` *page, click on the name of the workflow of interest. You will enter the home page of the Workflow, which describes the main features of the workflow including each step of the process. Click on the green button at the top right,* `Import workflow`, *and the workflow will be imported to your workflows page*

Figure 7.53 *Press the* `Import` *button and a notification appears informing you that the import was successful. You can follow the prompts to start using the link, or return to the previous page*

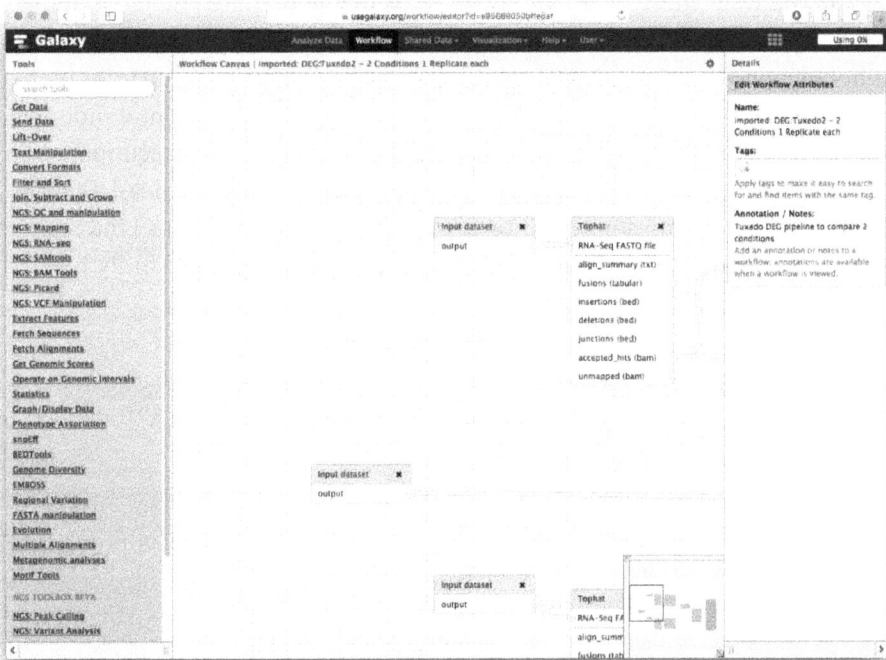

Figure 7.54 *Upon opening the recently imported research workflow, you can start modifying and customizing it to suit your needs. Whether it is done by adding your own workflow processes, or to delete unwanted parts, all content of the imported workflow on your own workflow panel is open for editing*

7.4 Installing and Setting Up Your Own Galaxy Server

Although public Galaxy servers are very convenient to use, there are still some inconvenient issues, for instance, uploading data that is too large, and the network speed is too slow because these public servers are in a foreign country. If you have a server on the internal local area network (LAN) running Galaxy, the speed of upload will obviously be much faster. A frequent problem is that the open server is far too slow, because the user load is typically too high, and the speed of completion of the work is always very slow. If you use your own private machine, you do not have this problem. Finally, the most important is the privacy issue. To compute highly sensitive information on a public Galaxy server, whose security is not assured, is the problem where this time, only using your own set of secured servers will guarantee the reliability you need.

Alternatively, these issues are superfluous to you because you just want to run Galaxy on your own Linux or Mac platform, with a web interface that you can use without having to enter complex commands on the command line in order to use these bioinformatics tools. It is possible to have Galaxy installed and mounted on your own machine — just make sure your computer has the execution environment of the Python programming language. You will need at least Python 2.5 before you can run most of the features of Galaxy. Now Mac OS or Linux environments (my example is Ubuntu) usually have a pre-installed built-in Python 2.5 or later versions.

Figure 7.55 *Verify that the operating system has installed a suitable version of Python using the terminal command,* `python --version`

Downloading the latest version of the Galaxy

Currently, for most large programming project development, there is a version control system, designed to ensure that for all jointly developed software, every amendment can be compared to previous releases. Each version of the same software can also be a candidate for further development; the complicated details of which involves software engineering professionals, which we will not discuss any further. Similarly, Galaxy also uses a version control system so that the user can download the latest officially maintained version. This official version control software server repository is called Git, which recommends using Ubuntu and other Linux OS for installation, which will be easier to install into than using macOS.

```
●  ○  ●           ⬆ eric — root@galaxy: ~ — ssh root@10.12.2.157 — 80×24
root@galaxy:~# git clone https://github.com/galaxyproject/galaxy/
The program 'git' is currently not installed.  You can install it by typing:
apt-get install git
root@galaxy:~#
```

Figure 7.56   *Enter* git clone https://github.com/galaxyproject/galaxy/. *From the official server, download the latest version of Galaxy. If you have not installed Git, you will be asked to install it. In Ubuntu, for example, please use the built-in package manager to install. The complete command is* sudo apt-get install mercurial

```
●  ○  ●           ⬆ eric — root@galaxy: /mnt — ssh root@10.12.2.157 — 80×24
root@galaxy:/mnt# git clone https://github.com/galaxyproject/galaxy/
Cloning into 'galaxy'...
remote: Counting objects: 192326, done.
remote: Compressing objects: 100% (239/239), done.
remote: Total 192326 (delta 123), reused 0 (delta 0), pack-reused 192085
Receiving objects: 100% (192326/192326), 61.31 MiB | 17.08 MiB/s, done.
Resolving deltas: 100% (152372/152372), done.
root@galaxy:/mnt#
```

Figure 7.57   *Once the Git installation is completed, you can start Galaxy download by repeating the command* git clone https://github.com/galaxyproject/galaxy. *All necessary files will be downloaded from the Github repository for Galaxy. The files are large, so depending on the speed of the network, it may take some time for the download to complete. Please wait until the command input cursor appears*

Starting your Galaxy server

In a nutshell, once the Galaxy download is complete, it is a single command install. All Galaxy integrated bioinformatics tools will be accessible. And there is a web server which will have to be running in order to allow you to use the browser to access them. So the first thing we need to do is to determine if the web server is really running.

```
● ○ ●          ⌂ eric — root@galaxy: /mnt/galaxy — ssh root@10.12.2.157 — 80×24
Initializing tool-data/quality_scores.loc from quality_scores.loc.sample
Initializing tool-data/regions.loc from regions.loc.sample
Initializing tool-data/sequence_index_base.loc from sequence_index_base.loc.samp
le
Initializing tool-data/sequence_index_color.loc from sequence_index_color.loc.sa
mple
Initializing tool-data/sift_db.loc from sift_db.loc.sample
Initializing tool-data/srma_index.loc from srma_index.loc.sample
Initializing tool-data/twobit.loc from twobit.loc.sample
Initializing static/welcome.html from welcome.html.sample
New python executable in .venv/bin/python
Installing setuptools, pip, wheel...done.
Activating virtualenv at /mnt/galaxy/.venv
Ignoring indexes: https://pypi.python.org/simple
/mnt/galaxy/.venv/local/lib/python2.7/site-packages/pip/_vendor/requests/package
s/urllib3/util/ssl_.py:90: InsecurePlatformWarning: A true SSLContext object is
not available. This prevents urllib3 from configuring SSL appropriately and may
cause certain SSL connections to fail. For more information, see https://urllib3
.readthedocs.org/en/latest/security.html#insecureplatformwarning.
  InsecurePlatformWarning
Collecting pip
  Downloading https://wheels.galaxyproject.org/packages/pip-8.0.0.dev0+gx1-py2.p
y3-none-any.whl (1.1MB)
    2% |█                            | 28kB 7.4MB/s eta 0:00:01
```

Figure 7.58   *Switch to the folder where the Galaxy kit resides. The default is that you execute* hg +x *command, to set the execute permissions. Then,* ./run.sh *to start loading all the relevant tools and configuration files. The first implementation requires a little bit more time*

```
● ○ ●          ⌂ eric — root@galaxy: /mnt/galaxy — ssh root@10.12.2.157 — 80×24
ddleware
galaxy.webapps.galaxy.buildapp DEBUG 2015-11-20 14:52:04,680 Enabling 'trans log
ger' middleware
galaxy.webapps.galaxy.buildapp DEBUG 2015-11-20 14:52:04,680 Enabling 'x-forward
ed-host' middleware
galaxy.webapps.galaxy.buildapp DEBUG 2015-11-20 14:52:04,681 Enabling 'Request I
D' middleware
galaxy.webapps.galaxy.buildapp DEBUG 2015-11-20 14:52:04,684 added url, path to
static middleware: /plugins/visualizations/charts/static, ./config/plugins/visua
lizations/charts/static
galaxy.webapps.galaxy.buildapp DEBUG 2015-11-20 14:52:04,684 added url, path to
static middleware: /plugins/visualizations/graphview/static, ./config/plugins/vi
sualizations/graphview/static
galaxy.webapps.galaxy.buildapp DEBUG 2015-11-20 14:52:04,685 added url, path to
static middleware: /plugins/visualizations/graphviz/static, ./config/plugins/vis
ualizations/graphviz/static
galaxy.webapps.galaxy.buildapp DEBUG 2015-11-20 14:52:04,685 added url, path to
static middleware: /plugins/visualizations/scatterplot/static, ./config/plugins/
visualizations/scatterplot/static
galaxy.queue_worker INFO 2015-11-20 14:52:04,685 Binding and starting galaxy con
trol worker for main
Starting server in PID 2077.
serving on http://127.0.0.1:8080
```

Figure 7.59   *When the output "*serving on http://127.0.0.1:8080*" appears, the Galaxy server is working. To end the Galaxy server process, you can use the* kill *command to delete the server's ID. The example here is PID 3490, as determined from the UNIX command,* ps -ef. *With the administrator privileges,* kill 3490, *you can terminate the Galaxy server. Alternatively, in the terminal window, you can also press* Ctrl + D, *to terminate the server*

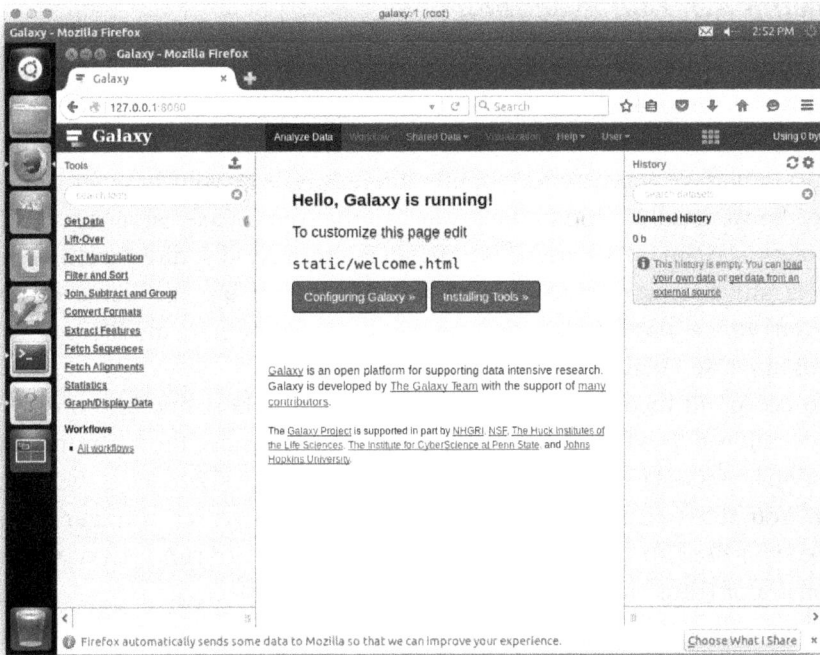

Figure 7.60 *With the Galaxy server running on your Linux machine, launch your browser, and go to the URL 127.0.0.1:8080, which shows the Galaxy Web interface listening on port 8080 (web server default is port 80)*

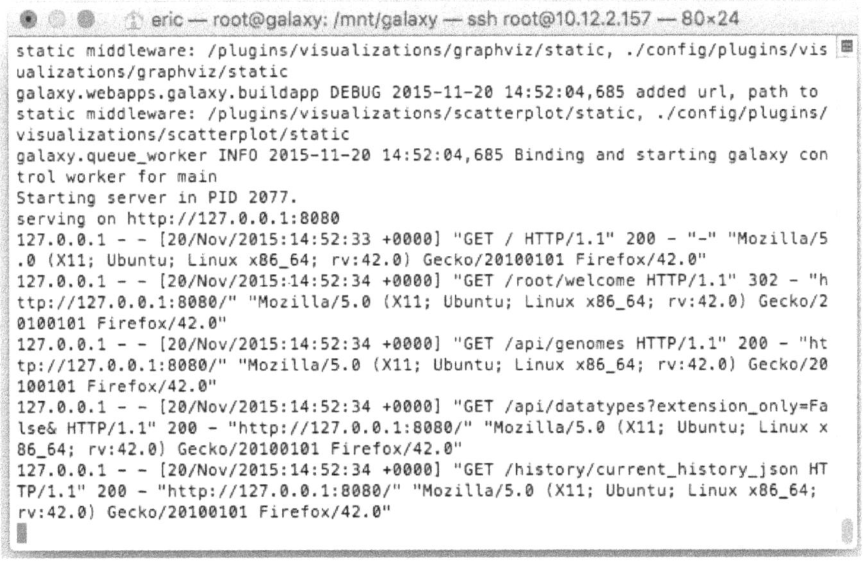

Figure 7.61 *While the Galaxy server is running, it will continue to update the screen to display all messages, such as which computer is connecting to it to run jobs*

Allowing external execution

When the Galaxy server starts up, the default is to only accept local connections, and only the computer it is installed in can access its resources. This is based on security considerations, but it is also impractical. If the server does not provide a graphical interface access to a computer, there is no way your browser can access Galaxy. So we need to modify the profile, so that the Galaxy server accepts other computers as clients. This profile is in the default galaxy-dist directory, and the file name is universe_wsgi.ini. Switch to this folder, and use the Linux command `nano universe_wsgi.ini` to use the nano text editor start modifying the contents of universe_wsgi.ini.

To accept all incoming IP addresses, modify the host line into host=0.0.0.0 (The default is `# host=0.0.0.0`, so in fact, you just remove the `#` comment sign). Once done, press `Ctrl + O` to save changes, press `Ctrl + X` to leave the nano text editor. You must remember that every time Galaxy server configurations are revised, you need to restart the Galaxy server (that is, re-run Galaxy `run.sh` as administrator (root user) for the new settings to take effect.

When Galaxy restarts, it will now accept all Internet addresses connecting to it and the server screen will show "`serving on 0.0.0.0:8080`" if you are on the server viewing it at `http://127.0.0.1:8080`. This means that any client browser from any computer from any IP address can now remotely operate the Galaxy server, so long as both the client computer and the server are on the Internet. The client of course needs to know what is the Galaxy server's IP address, e.g. 123.234.56.78. Then entering the URL http://123.234.56.78:8080 on the client browser will get the user to the Galaxy server you have just set up. Even better, you can get a domain name, say from no-ip.org and match that to your server IP address, say http://mygalaxy.no-ip.org and that will work too.

Installation of bioinformatics tools

I would like to stress once again: Galaxy is just a graphical interface to help you to call bioinformatics tools from bioinformatics software packages! So it will not contain all these third party tools. Besides loading the entire suite of software, which may not be suitable for everyone, you will need a huge storage capacity. So, if you wish to use a particular bioinformatics tool through the Galaxy interface, you have to manually install it. So what tools should you install? Consult the Galaxy Wiki. For high-throughput sequencing studies, the most commonly used sequence alignment tool is Bowtie, which would have to be downloaded to install and interface with Galaxy. After installation, to be successfully called by Galaxy can be a little tricky.

```
● ● ●    ⌂ eric — root@galaxy: /mnt/galaxy/config — ssh root@10.12.2.157 — 80×24
  GNU nano 2.2.6                 File: galaxy.ini                 Modified  ▦

[server:main]

# The internal HTTP server to use.  Currently only Paste is provided.  This
# option is required.
use = egg:Paste#http

# The port on which to listen.
#port = 8080

# The address on which to listen.  By default, only listen to localhost (Galaxy
# will not be accessible over the network).  Use '0.0.0.0' to listen on all
# available network interfaces.
host = 0.0.0.0▊

# Use a threadpool for the web server instead of creating a thread for each
# request.
use_threadpool = True

# Number of threads in the web server thread pool.

^G Get Help  ^O WriteOut  ^R Read File  ^Y Prev Page  ^K Cut Text   ^C Cur Pos
^X Exit      ^J Justify   ^W Where Is   ^V Next Page  ^U UnCut Text ^T To Spell
```

Figure 7.62 *With nano editor to edit the host line in the configuration file, and modify it into* host = 0.0.0.0, *and thereafter* Ctrl + O *and* Ctrl + X *to save and exit*

```
● ● ●    ⌂ eric — root@galaxy: /mnt/galaxy — ssh root@10.12.2.157 — 80×24
ddleware                                                                    ▦
galaxy.webapps.galaxy.buildapp DEBUG 2015-11-20 15:07:45,732 Enabling 'trans log
ger' middleware
galaxy.webapps.galaxy.buildapp DEBUG 2015-11-20 15:07:45,732 Enabling 'x-forward
ed-host' middleware
galaxy.webapps.galaxy.buildapp DEBUG 2015-11-20 15:07:45,732 Enabling 'Request I
D' middleware
galaxy.webapps.galaxy.buildapp DEBUG 2015-11-20 15:07:45,735 added url, path to
static middleware: /plugins/visualizations/charts/static, ./config/plugins/visua
lizations/charts/static
galaxy.webapps.galaxy.buildapp DEBUG 2015-11-20 15:07:45,735 added url, path to
static middleware: /plugins/visualizations/graphview/static, ./config/plugins/vi
sualizations/graphview/static
galaxy.webapps.galaxy.buildapp DEBUG 2015-11-20 15:07:45,736 added url, path to
static middleware: /plugins/visualizations/graphviz/static, ./config/plugins/vis
ualizations/graphviz/static
galaxy.webapps.galaxy.buildapp DEBUG 2015-11-20 15:07:45,736 added url, path to
static middleware: /plugins/visualizations/scatterplot/static, ./config/plugins/
visualizations/scatterplot/static
galaxy.queue_worker INFO 2015-11-20 15:07:45,736 Binding and starting galaxy con
trol worker for main
Starting server in PID 2260.
serving on 0.0.0.0:8080 view at http://127.0.0.1:8080
▊
```

Figure 7.63 *Galaxy accepts any remote connection server. The debug screen prompt message becomes* "serving on 0.0.0.0:8080", *which is viewable from your server at http://127.0.0.1:8080*

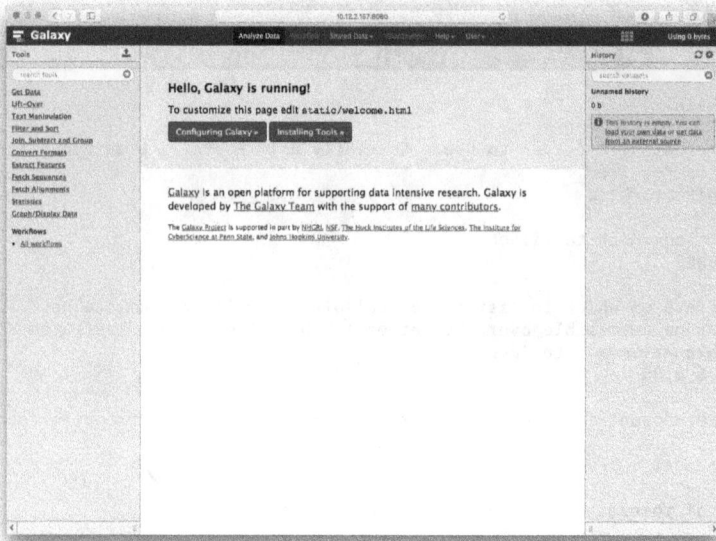

Figure 7.64   In my example, the Linux server installed has the LAN IP address 10.12.2.157, which is a private network address space. Only other computers within the local area network (LAN) can use this address in their browser; by entering http://10.12.2.157:8080 one can see the Galaxy screen. Unless your IP address is a public address, your users outside from the Internet cannot access this Galaxy server

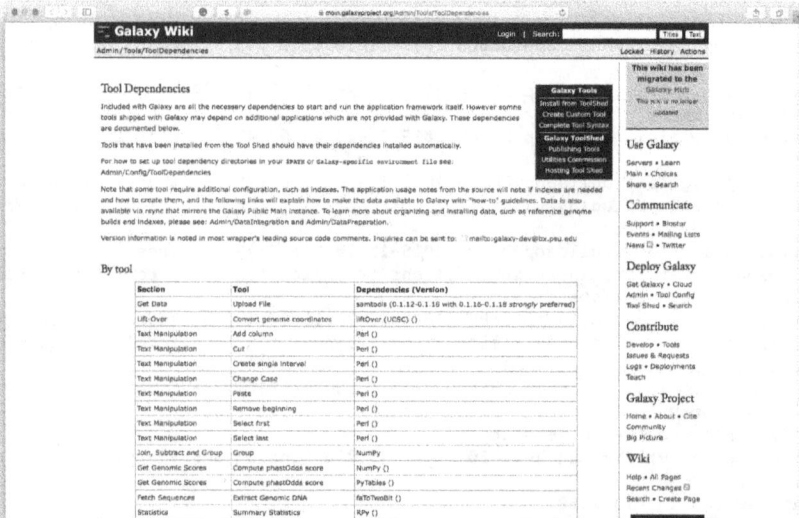

Figure 7.65   Galaxy Wiki page Tool Dependencies (https://moin.galaxyproject.org/Admin/Tools/ ToolDependencies). This lists all the tool dependencies, i.e. what programs need to be pre-installed in order to make the software packages that you want to work properly

In previous Galaxy releases, the user or administrator has to install the bioinformatics related tools manually. Download the correct tool which is necessary, install it and register it to the Galaxy configuration file. If any step is incorrect, the installed tool cannot be executed *via* Galaxy web interfaces. Luckily, Galaxy has its own "package management" system now (also known as "Tool Shed"). Similar to the Ubuntu's apt-get or Red Hat's yum, most of the bioinformatics toolchains can be installed by auto. All you have to do is enable the administrator account by modifying the configuration file.

```
⦿ ◎ ⦾      🏠 eric — root@galaxy: /mnt/galaxy — ssh root@10.12.2.157 — 80×24
   GNU nano 2.2.6             File: config/galaxy.ini              Modified   ▦

 # addresses or user names being passed to Galaxy - set the following option
 # to True to force these to lower case.
 #normalize_remote_user_email = False

 # Administrative users - set this to a comma-separated list of valid Galaxy
 # users (email addresses).  These users will have access to the Admin section
 # of the server, and will have access to create users, groups, roles,
 # libraries, and more.  For more information, see:
 #    https://wiki.galaxyproject.org/Admin/Interface
 admin_users = leezhenyu@gmail.com

 # Force everyone to log in (disable anonymous access).
 #require_login = False

 # Allow unregistered users to create new accounts (otherwise, they will have to
 # be created by an admin).
 #allow_user_creation = True

 # Allow administrators to delete accounts.

^G Get Help    ^O WriteOut   ^R Read File  ^Y Prev Page  ^K Cut Text   ^C Cur Pos
^X Exit        ^J Justify    ^W Where Is   ^V Next Page  ^U UnCut Text ^T To Spell
```

Figure 7.66   *To enable the administration privileges, please edit the galaxy.ini file, which is inside the config folder. Find the line which starts with "admin_user", after the equal sign, append the email address of administrator. If you need more than one administrator, please use comma as separator, append other administrator email accounts. Do not forget to save the file after modification*

```
●  ◉  ●      🔒 eric — root@galaxy: /mnt/galaxy — ssh root@10.12.2.157 — 80×24
[root@galaxy:/mnt/galaxy# mkdir tool_dep
[root@galaxy:/mnt/galaxy# chmod 755 tool_dep/
root@galaxy:/mnt/galaxy# █
```

Figure 7.67 *Create a folder for tool installation, Tool Shed will use it. Now I created a "tool_dep" folder under galaxy root folder, and assigned the permission number 755 by* chmod

```
●  ◉  ●      🔒 eric — root@galaxy: /mnt/galaxy — ssh root@10.12.2.157 — 80×24
  GNU nano 2.2.6            File: config/galaxy.ini                 Modified

# <toolbox> tag.
#tool_path = tools

# Path to the directory in which tool dependencies are placed.  This is used by
# the tool shed to install dependencies and can also be used by administrators
# to manually install or link to dependencies.  For details, see:
#    https://wiki.galaxyproject.org/Admin/Config/ToolDependencies
# If this option is not set to a valid path, installing tools with dependencies
# from the Tool Shed will fail.
tool_dependency_dir = /mnt/galaxy/tool_dep█

# File containing the Galaxy Tool Sheds that should be made available to
# install from in the admin interface (.sample used if default does not exist).
#tool_sheds_config_file = config/tool_sheds_conf.xml

# Set to True to enable monitoring of tools and tool directories
# listed in any tool config file specified in tool_config_file option.
# If changes are found, tools are automatically reloaded.  Watchdog (
# https://pypi.python.org/pypi/watchdog ) must be installed and

^G Get Help  ^O WriteOut   ^R Read File ^Y Prev Page ^K Cut Text  ^C Cur Pos
^X Exit      ^J Justify    ^W Where Is  ^V Next Page ^U UnCut Text^T To Spell
```

Figure 7.68 *Re-open the galaxy.ini file, find the line that starts with* "tool_dependency_dir", *append the created tool installation path after the equal sign. Save the updated configuration file and restart the Galaxy server*

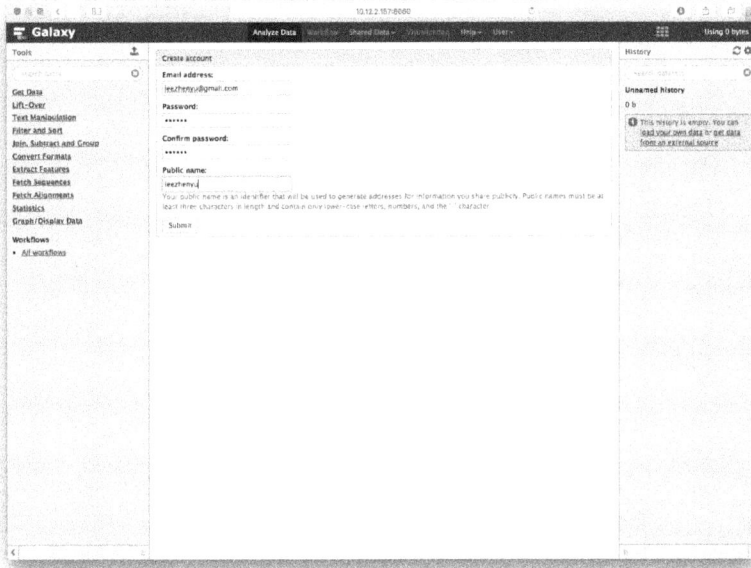

Figure 7.69 We just enabled the administrator function, so it is time to add an administrator. Open the Galaxy portal by browser, register a new account, please remember to use the email address which is recorded in the galaxy.ini file

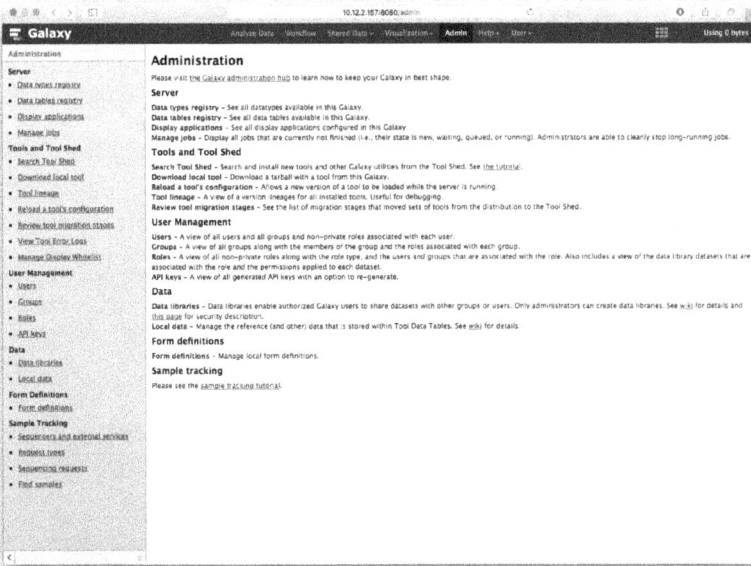

Figure 7.70 Now login to the Galaxy portal with the registered administrator email, you should see an "Admin" item on the top, please click it

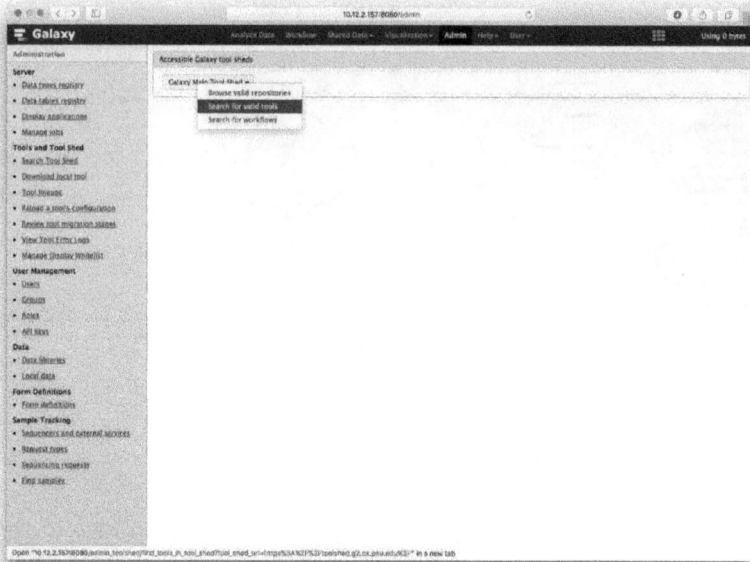

Figure 7.71 *On the Administrator page, there is a "*Search Tool Shed*" item on the left menu, click it. And you will see a "*Galaxy Main Tool Shed*", click it and choose "*Search for valid tools*". Or click "*Browse valid repositories*" to check all supported bioinformatics tools*

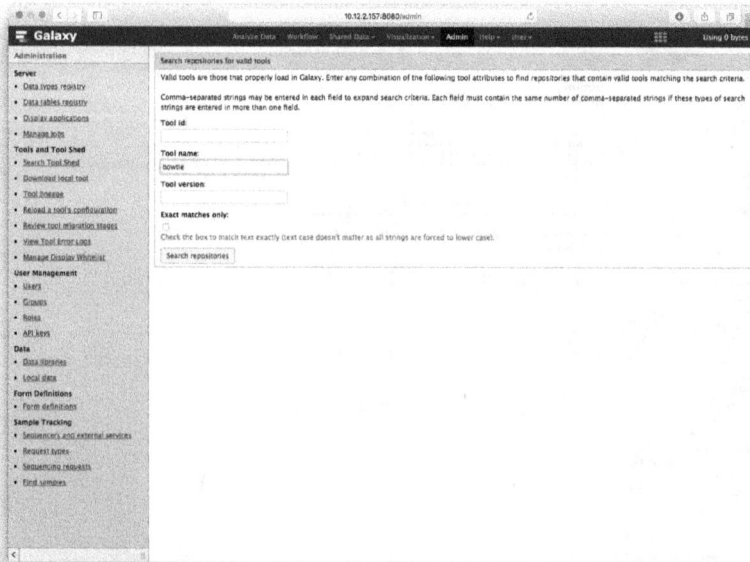

Figure 7.72 *Type "bowtie" or "bowtie2" in the* Tool name *field and click the "*Search repositories*" button*

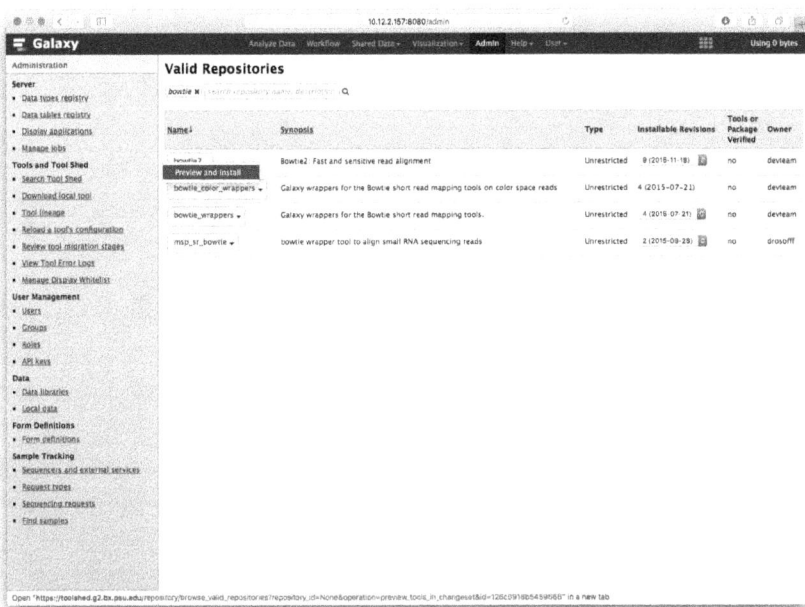

Figure 7.73 Click the bowtie2 button first, and then click "Preview and install"

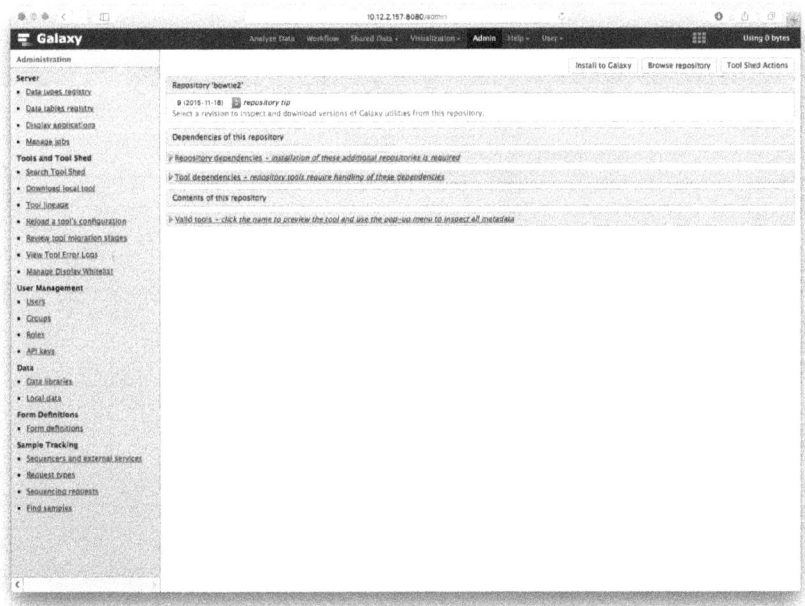

Figure 7.74 Confirm the version for installation, click the "Install to Galaxy" button on the upper-right

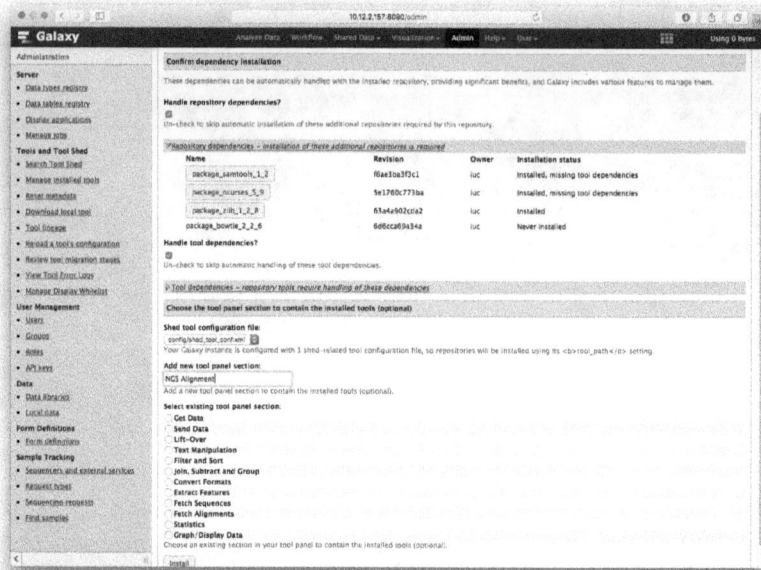

Figure 7.75 *Show the details of installation packages and its dependencies. Please also check the* "Panel Section", *it is the category for the new installed tool. You can choose default category from* "Select existing tool panel section" *or* "Add new tool panel section" *and provide the category name (e.g. NGS Alignment)*

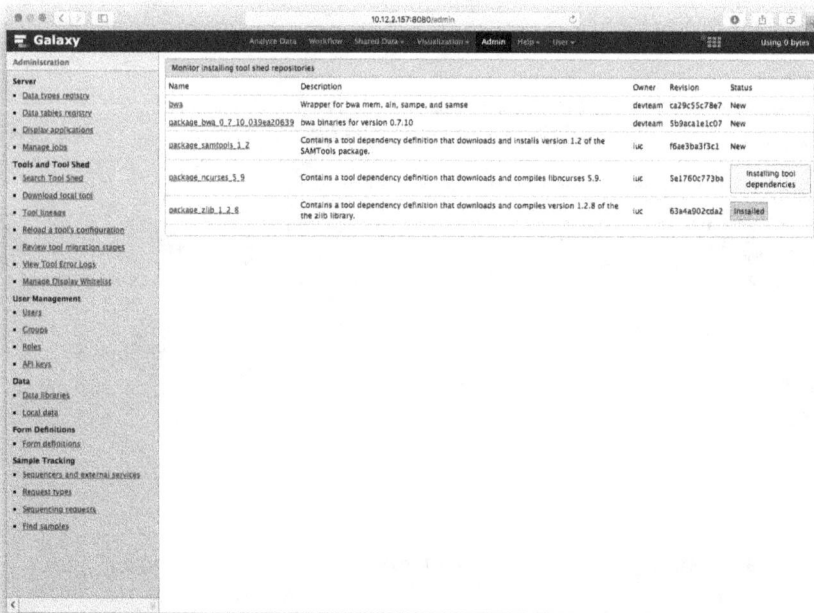

Figure 7.76 *When you start the tool and dependencies installation, you can see the progress*

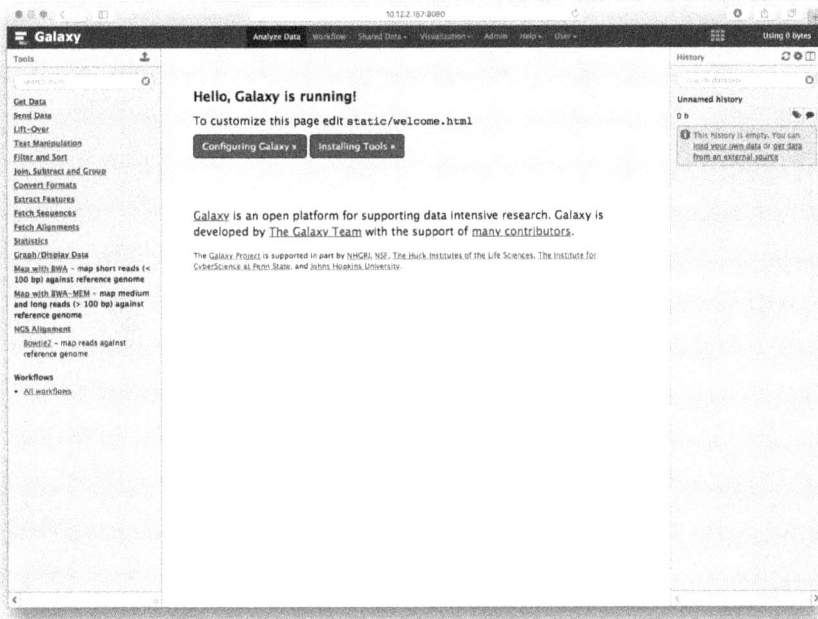

Figure 7.77 *After installation, there is a new category* "NGS Alignment" *in the toolchain menu,* "Bowtie2" *inside it*

Adding new reference sequences

Almost every bioinformatics tool, especially sequence alignment tools, need to access reference sequence data. Since we are using the same set of data, we can store them on a particular fixed path in the hard disk, so that bioinformatics tools or Galaxy can access them. Here we demonstrate two tools to add a reference sequence and to carry out profile editing. Galaxy does not install reference sequences by default. It also needs to use the profile manager to indicate where the reference sequence is kept.

Latest Galaxy build has data management feature, including the management of reference sequences. With the "Data Management" package, you will see the detail of data. Now we demonstrate the usage of Data Management package, use the configuration of Bowtie2 reference sequence as example.

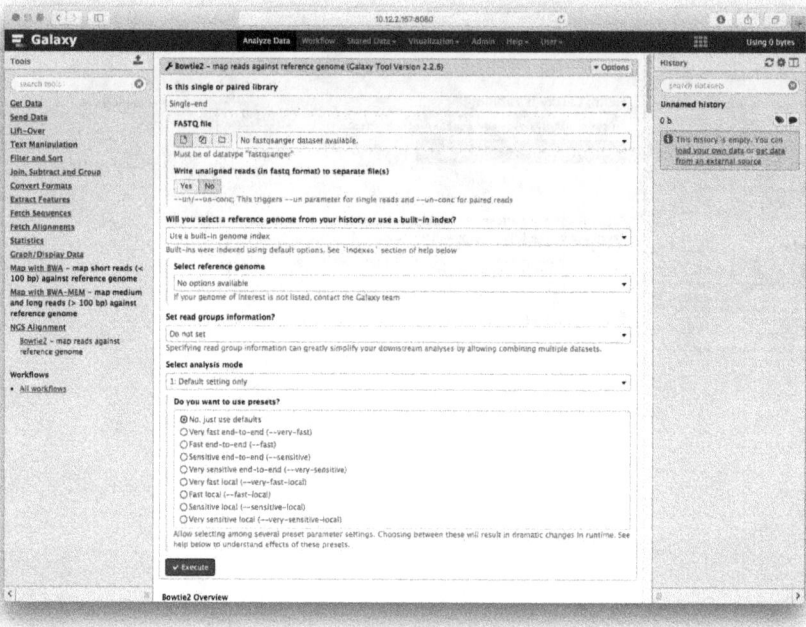

Figure 7.78 *Try to use Bowtie2, but unable to choose any reference sequence? The root cause is you did not install and configure it*

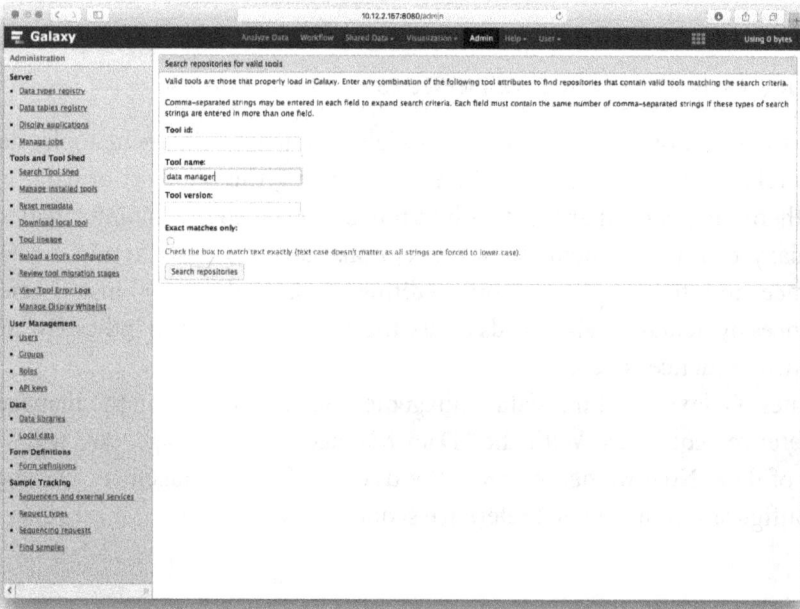

Figure 7.79 *Use Tool Shed again, search the "data manager" in the tool name*

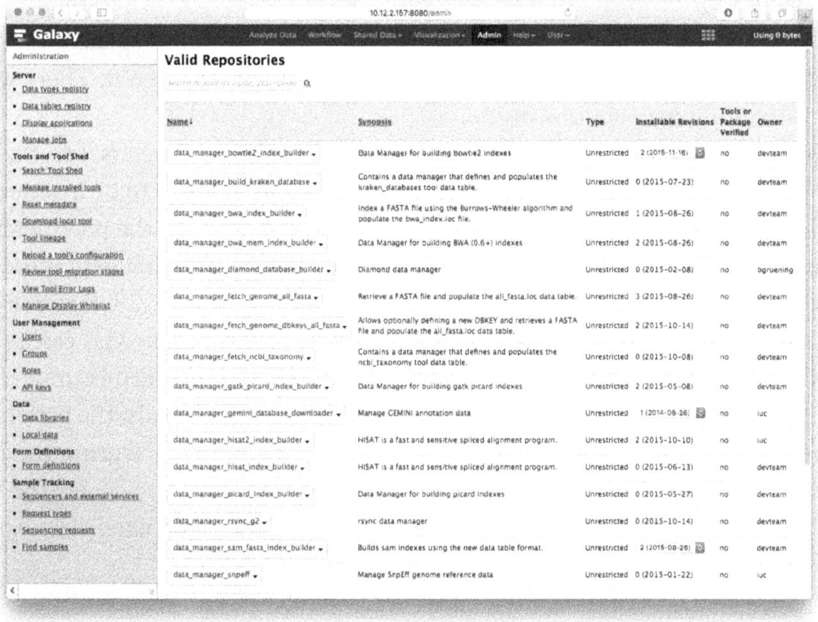

Figure 7.80 There is a search result; "`data_manager_boetie2_index_builder`". Please install it

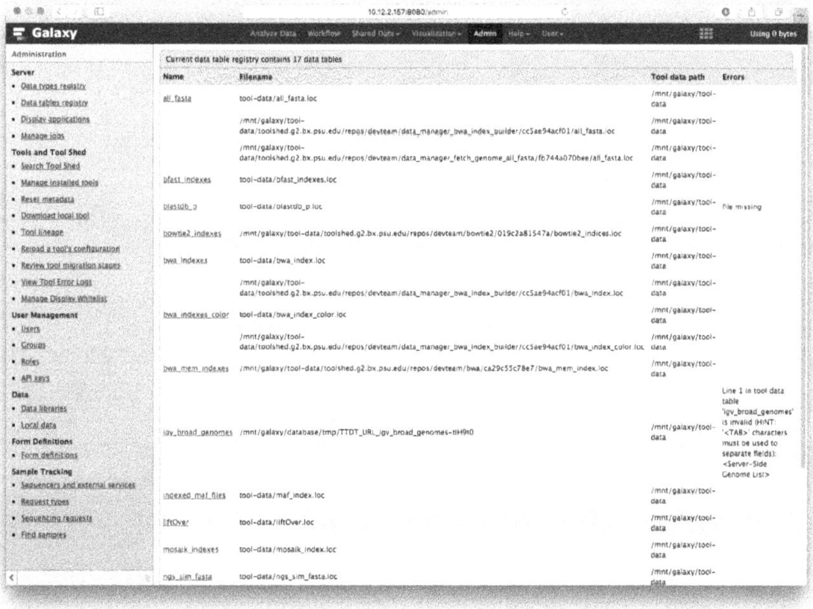

Figure 7.81 Click the "`Data table registry`" on left hand side menu, you can see the location for installed tools related data and its full path. For example, the path for installed Bowtie2 is /mnt/galaxy/ tool-data/... (detail as follows)

Galaxy tool profiles are found in the galaxy-dist folder, inside tool-data for each tool has a profile. For example, for Bowtie2, its profile is galaxy-dist/tool-data/bowtie2_indices.loc. Please use the nano text editor to open it. There will be basic formatting instructions inside.

For the hg19 example, there are a total of four parameters for the line: hg19 hg19 Human (hg19) /refidx/hg19/bowtie_path/base/hg19 to be inserted into bowtie2_indices.loc. Inside this file, ensure that space between the parameters are Tab Separated, you cannot use space characters as blanks. The first field is the hg19 version identification ID (Unique build ID), the second is the hg19 database identification value (if you want to accelerate Galaxy's performance, you can refer to the official Wiki to study how to establish a database), and the third Human (hg19) field is the value to be displayed in the Galaxy menu name. Finally, the path is very important, remember to complete the full path to hg19 folder. During the Linux operation of Bowtie2, these parameters are not only necessary to access the reference sequence hg19, they are needed for other index files with extensions such as .bt2 files.

```
●  ○  ●    ⌂  eric — root@galaxy: /mnt/galaxy/tool-data — ssh root@10.12.2.157 — 80×24
  GNU nano 2.2.6              File: bowtie2_indices.loc

#      -rw-rw-r-- 1 james    james    914M Feb 10 18:56 hg19canon.1.bt2
#      -rw-rw-r-- 1 james    james    683M Feb 10 18:56 hg19canon.2.bt2
#      -rw-rw-r-- 1 james    james    3.3K Feb 10 16:54 hg19canon.3.bt2
#      -rw-rw-r-- 1 james    james    683M Feb 10 16:54 hg19canon.4.bt2
#      -rw-rw-r-- 1 james    james    914M Feb 10 20:45 hg19canon.rev.1.bt2
#      -rw-rw-r-- 1 james    james    683M Feb 10 20:45 hg19canon.rev.2.bt2
#
# then the bowtie2_indices.loc entry could look like this:
#
#hg19    hg19    Human (hg19)      /depot/data2/galaxy/hg19/bowtie2/hg19canon
hg19     hg19    Human (hg19)      /data/refIdx/bowtie2/hg19/hg19
#
#More examples:
#
#mm10    mm10    Mouse (mm10)      /depot/data2/galaxy/mm10/bowtie2/mm10
#dm3     dm3           D. melanogaster (dm3)   /depot/data2/galaxy/mm10/bowtie$
#
#

^G Get Help   ^O WriteOut   ^R Read File  ^Y Prev Page  ^K Cut Text   ^C Cur Pos
^X Exit       ^J Justify    ^W Where Is   ^V Next Page  ^U UnCut Text ^T To Spell
```

Figure 7.82 *Use the nano text editor to edit galaxy-dist/tool-data/bowtie2_indices.loc, write the full path of hg19 reference sequence. This is the Bowtie2 profile, and finally remember to press* Ctrl + O *to save and* Ctrl + X *to exit*

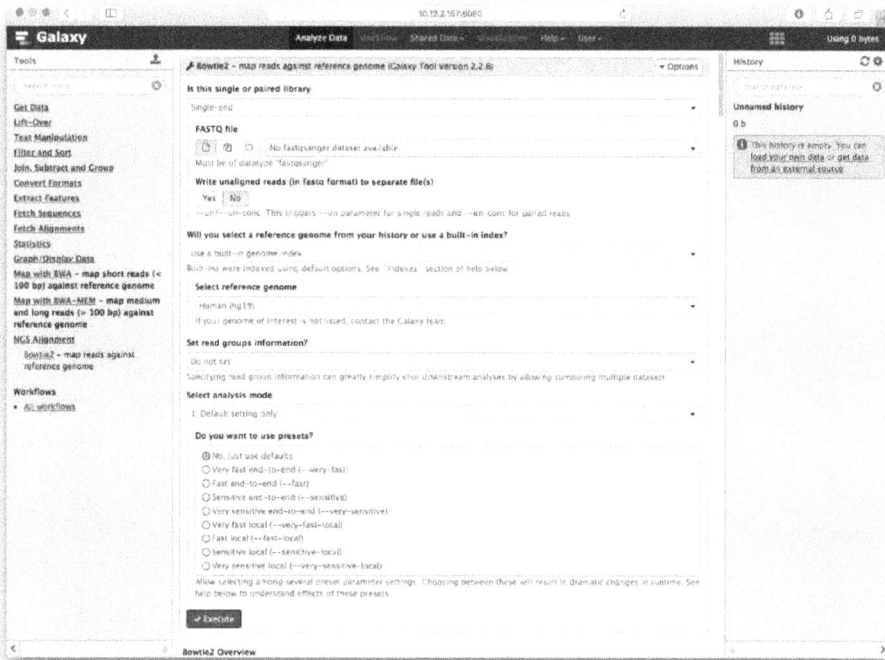

Figure 7.83 *Having edited bowtie2_indices.loc, save it and restart the Galaxy server. From Bowtie2 menu, you will now see the newly added hg19 reference sequence*

In this section, we have learnt how to create a simple internal Galaxy server and together with useful bioinformatics tools and the necessary reference sequence datasets, shown how useful they can be. As this book is not a Galaxy manual, we will not explain other important settings, other than features suitable for a beginner's level. Interested readers and advanced users are strongly recommended to explore this powerful interface, and conduct your own in-depth study of the Galaxy Wiki.

APPENDIX

Learning Regular Expressions through Practising Simple Data Processing

In today's bioinformatics research world, the biggest challenge is to process the data deluge coming from ever increasingly high-throughput, Next Generation Sequencing machines. As a result of the data being only in the form of ATCG, bioinformatics research has now shifted to a "string handling" juggle.

Biological raw data, after conversion into text characters, are processed through a variety of algorithms to make efficient string manipulation, so as to extract biological meaning from them. Since bioinformatics data are very large, it is impossible to use manual methods to do this; only ever-increasingly powerful computers and faster algorithms can analyze the data. In this section, for beginners, we stress the need for a good appreciation for understanding the concept of regular expressions and how they might be used for simple data processing.

Regular Expressions

Suppose you have a file full of text characters, and you only want to select specific lines. To get these lines, you need to find out a specific string of characters which is unique to these lines and not found in others. There are several programs you can use to do this. The most popular is grep. It is so popular, that it has become a verb, like "let's grep the string from this file", very much like the verb "to google" for a keyword. These string search programs invariably rely on a search pattern language called regular expressions. We will take you on a simple tour of using examples to understand regular expressions (or regex) and how useful as a tool they are to next generation sequencing.

Say you have a file containing a set of simple FASTA sequences. You want to know how many sequences you have, each of which by definition of FASTA format, has a header line beginning with ">" in a single line preceding the sequence lines.

One character pattern match

Let's grep for the headers. Theoretically and by definition, the only commonality amongst all headers in a FASTA formated file are ">" beginning the line, and sequences do not contain this character.

```
grep ">" sequences.fasta
```

Here, the command grep takes in ">" and `sequences.fasta` as two valid arguments. ">" is the string pattern at its simplest, just one character. `sequences.fasta` is the filename which you want to grep to scan through.

So this is called the literal character pattern matching. You can match all kinds of characters as supported by various programs, such as grep, sed, awk and programming languages, such as Perl. Literal characters match literally the characters they represent. What about characters which are non-literal? In regex, they have a special meaning, and they are called metacharacters. We will discuss them soon, but first, let us familiarize ourselves by practising some useful commands with the grep utility program.

Numbering a file and printing line number of a hit

This command will produce all header lines, as lines matching the character ">". You can number the lines of the file and print the line number e.g.:

```
grep -n ">" sequences.fasta
```

Here, in addition to ">" and `sequences.fasta`, we have -n, which is a special flag, which tells grep to switch on line counting and printing. So grep takes on flags such as -n to instruct it to do something different from its default basic action, in addition to the input file, and the regex itself.

Note that the regex is enclosed within double quotes just to make sure with all these other flags and characters around, we unambiguously inform grep, where the regex is located, i.e. within the double quotes.

Counting number of grepped hits

If you use the -c flag, you can count the number of lines instead of printing them out.

```
grep -c ">" sequences.fasta
```

UNIX redirection using pipes

There is a longer way:

```
grep ">" sequences.fasta | wc -l
```

Pipe is the unix redirector, which pipes the standard output of grep into the standard input of another program called wc (word count). It sends the stream of output from grep into the standard input of the wc program, which counts words. But if you set wc flag -l, then wc will count l for lines, and return the total number of lines. This is the powerful feature of UNIX, which allows redirection of output and input.

Grep and output several lines of context around the hit

Just in case you may have sequence headers with no sequence, maybe you want to print the context around the grep. This is possible.

```
grep -C 1 ">" sequences.fasta
```

This -C (for context) accepts an argument, which is a number, and here it means print 1 line of the context of the match, i.e. one line before (-B) and one line after (-A) the matched line. Each match is separated from the other by two hyphens. So you can visually check if two or more consecutive lines are header lines.

Grepping for non-matching lines

What if you wanted sequences only, and not the headers?

```
grep -v ">" sequences.fasta
```

The -v flag instructs grep to give us non-matches, in this case, non-header lines.

Grepping for unwanted characters

What if we wanted to check if there were other characters in the sequence other than A, T, C and G. This is an important task because sometimes data are corrupted, or contain unexpected characters, which we did not anticipate. This assumption can lead to errors in programs downstream processing the "bad" data.

Mistake of logic

We could take the -v flag to generate the sequences, and then we could search for A, T, C or G in every line and report those lines that did not match these characters again, using the -v flag. You may think we could use the UNIX pipe (|) here to good effect. So one may make the following mistake:

```
grep -v ">" sequences.fasta | grep -v "A" | grep -v
"T" | grep -v "C" | grep -v "G"
```

So this means that do grep on the sequences.fasta file, and output all lines not containing ">", i.e. only sequence lines will be outputted and sent to the next grep command, which is to search for all A, and output all lines that do not contain As and send it to the next grep. The next grep searches for T and -v flag outputs all those that do not contain Ts, and so on for non-C and non-G lines.

This is a mistake of logic. For example, some lines containing A, may comprise non-T, non-C and non-G characters. So we need to use another way of finding lines that comprise non-A, non-T, non-C and non-G characters all in one go. This series of greps actually says, grep search for lines that do not have ">", i.e. the sequence lines only, and then for each of these lines, match for A and if they have A somewhere, it is a match, do not send it to the standard output (stdout); if it does not match an A anywhere on that line, send it to stdout. Of the lines which do not comprise a single A, grep for Ts, and output those lines that do not comprise even a single T... and so on. This series of greps actually outputs lines that do not have an > and not a single A and not a single T and not a single C and not a single G, which is not what we were asking for. We were asking for lines which may comprise non-ATCG characters mixed with ATCGs.

In other words, for every character in each line that grep examines, we need to ask all four questions at the same time, is it A, T, C, G and if not, we ask the same four questions of the next character.

So obviously this command, fancy as it may be, will not work for us. We need to somehow inform grep, we are looking for ATCGs and no other characters.

Egrep or grep -E extended regular expression grep

Here, we are basically saying to grep, please match, A or T or C or G. For this, we need to extend grep to be able to handle the OR operator, which grep uses as "|" the same pipe character. This can only be handled by egrep or using the grep with the -E (for extended regular expression) flag.

```
egrep "A|T|C|G" sequences.fasta
```
or
```
grep -E "A|T|C|G" sequences.fasta
```

Here the | in grep is different from the UNIX redirector pipe. Please note the difference. The three pipes | are enclosed in double quotes "A|T|C|G" and fed into grep as an argument. It is thus protected from the bash interpreter from interpreting the | as a UNIX redirector command.

So if grep "A|T|C|G" sequences.fasta will generate lines which contain ATCGs, we need to find lines that do not contain ATCGs. You may think ok, let us give it a -v flag. Find all lines that do not contain ATCGs?

```
grep -E -v "A|T|C|G" sequences.fasta
```

This does exactly what the previous series of greps does. It is also a mistake. It only shows lines that do not contain ATCGs at all. But what if there was a line with ATCGs and other characters? Just as before, this command will not pick it up! It is in fact, equivalent to the earlier series of greps.

Egrep and the character class

What we need is a way to find non-ATCG characters. We can use the regular expression — character class. This involves putting all characters you wish to match in a class, which you can create by placing them in square brackets. Note that here, we are not matching the [or]. We are merely using them as part of the regular expression syntax to denote a special meaning, i.e. anything enclosed between them is a character class of literal characters at that position, which can be successfully matched to the character in the string.

```
grep "[ATCG]" sequences.fasta
```

This means for one character position as denoted by [ ], accept matches either A or T or C or G. This is totally different from matching a string "ATCG", i.e. consecutively A followed by T followed by C and by G.

This picks up all lines including sequence lines which match A or T or C or G. If we set the flag -v, will we get what we need? No!

```
grep -v "[ATCG]" sequences.fasta
```

will only get us the lines that do not comprise a single A or T or C or G! What about those lines that comprise something else mixed with A or T or C or G?

Meta characters, their special meanings and matching their literal, non-metacharacter meaning using backslash. Note that in regex, [and] have this special meaning. They belong to a group of special characters which have a special meaning in regex. They are called metacharacters, and include 12 characters: the backslash \, the caret ^, the dollar sign $, the period or dot ., the vertical bar or pipe symbol |, the question mark ?, the asterisk or star *, the plus sign +, the opening parenthesis (, the closing parenthesis), the opening square bracket [, and the opening curly brace {.

If we want to match one of these characters, we have to use backslash as a prefix, including backslash itself. For example, we want to match \ itself, we have to use a backslash of a backslash \\ in order to match a single backslash in its non-special meaning literal character sense. So if you wish to match any of these special characters, precede it with a backslash.

Egrep character class negation

The solution to the above question is to use negation special character, the caret ^.

```
grep "[^ATCG]" sequences.fasta
```

Therefore inside the class of characters, by placing a hat or caret character ^ in front of the letters, we negate them. We are saying, for each character that grep is matching, test if it is not A or not T or not C or not G. So long as there is a match to a non-A or non-T or non-C or non-G character somewhere in the line, that is a match. Print them out.

Remember that caret ^ is also a special meta character, and that can be easily done to make it literal of we are using a :

Obviously, > is a non-A or non-T or non-C or non-G character. Let us filter that out.

```
grep "[^ATCG]" sequences.fasta | grep -v ">"
```

This command says, find all lines in `sequences.fasta` that contain a non-A or non-T or non-C or non-G character, and then filter out those lines that do not contain a >, i.e. forget about these lines that are headers.

Regular expression: Beginning of line anchor ^

But what if > occurs somewhere in a sequence line? That is not a header line! Header lines have lines beginning with ">". So how do we match a line having ">" right at the beginning?

```
grep "[^ATCG]" sequences.fasta | grep -v "^>"
```

Again, we use the caret or hat character. This time it means something totally different. Outside the [] character class, ^ means "match beginning of a line". Because there is a shortage of such characters, called meta-characters like ^, they are reused in different contexts; when you are in UNIX, you have to be very careful about this.

So grep -v "^>" means match for lines beginning with >, and the -v says, print the lines that do not satisfy this pattern.

So -v refers to the line, ^ inside [^ATCG] refers to the character in the line and the ^ in ^> anchors the > pattern to the beginning of the line.

Case-sensitive and case-insensitive grep

So what if we have atcg in lower case. Because UNIX is case sensitive, it will not match ATCG, so we have to state them as [^ATCGatcg] or use the -i flag, which is the case-insensitive flag

```
grep "[^ATCGatcg]" sequences.fasta | grep -v "^>"
```
or
```
grep -i "[^ATCG]" sequences.fasta | grep -v "^>"
```

This will pick out all sequence non-header lines, and check if they comprise any non-A or non-T or non-C or non-G characters.

So if there are Ns, it will be printed. If there are blank space characters embedded in a sequence or at the beginning of a line, or trailing a line, they will be picked up.

Regular expression: End of line anchor

So if you have in your FASTA files lines which contain blank spaces at the end of the line, which you did not see. And you want to check if there are lines like that in your FASTA files, called trailing blank spaces. You can search for them by looking for blank space anchored at the end of the line.

```
grep " $" sequences.fasta
```

$ is the end of line anchor. It represents end of the line just like ^ represents the beginning of a line. In other words, in " $" the blank space in front of the $ indicates that the blank space is right next to the end of the line anchor, i.e. it is a trailing blank space.

These are very useful commands for you to remember, to check up on lines in your FASTA files, in case there are some funny characters that creep into your data.

ALWAYS CHECK YOUR DATA FOR BAD DATA!

Many programs do not check for bad data and behave in an unpredictable manner when unexpected input is received. It is our job as users to make sure the assumptions made by the programmer regarding the input are correct.

More about regular expressions

If you have a nucleotide sequence file in FASTA format, and you are searching for a sequence pattern, say a Restriction Enzyme site GAATTC (EcoRI restriction site).

First let us get the FASTA sequence lines

```
grep -v "^>" dna.fasta
```

Then we need to join them all up in one single line

```
grep -v "^>" dna.fasta | tr -d "\n"
```

tr is the UNIX translate command to change characters from one to another. tr -d is the flag that deletes any matching character. "\n" is the newline character. Therefore, tr -d "\n" is to delete all newline characters. Any input stream containing a newline character will be deleted, resulting in the concatenation of every line into a single very long line.

From this output stream of a single line of sequence data, we can do a search for the sequence pattern GAATTC.

```
grep -v "^>" dna.fasta | tr -d "\n" | grep -i "GATTCC"
```

GAATTC is the string we wish to match. -i flag makes it case-insensitive to take care of any sequence input that may have mixed upper and lower cases.

But this does not quite help us find the location. So let us invoke the coloring flag for grep. -color=auto will automatically colour the matching string with a contrasting colour against the background.

```
grep -v "^>" dna.fasta | tr -d "\n" | grep -i
-color=auto "GAATTC"
```

But a DNA sequence is typically double-stranded. So we need to search the opposite strand too. However, in this case we do not need to because this is a DNA palindrome. The DNA sequence is directional, and is rendered 5' to 3' by convention. The other strand, which is called the complementary strand looks like this

```
5'-GAATTC-3'
3'-CTTAAG-5'
```

So, the opposite strand, when rendered by convention in the 5' to 3' direction, will still be GAATTC.

```
3'-CTTAAG-5' is 5'-GAATTC-3'
```

5'-CTTAAG-3' is a totally different sequence from 5'-GAATTC-3' and is a different restriction enzyme that recognizes this sequence.

```
grep -v "^>" dna.fasta | tr -d "\n" | grep -i
-color=always "GAATTC"
```

-color=always, means that grep will keep the colour flags intact when the output is generated, not just for the UNIX terminal screen (which what auto does, it removes the colour tags when saving the output).

Since the colour flags are kept intact using -color=always, we can further process the output and still retain the colour.

```
grep -v "^>" dna.fasta | tr -d "\n" | grep -i
-color=always "GAATTC" | fold -w 60
```

By piping the output to the fold command, we can wrap the long lines into 60 characters per line. However, this unfortunately includes the colour flag characters, which total about 17 characters including various control characters that instruct the UNIX terminal to change colour on and off before and after the matching strings respectively! But it gives you a rough idea of what grep can do for you.

Even more regular expression

Meta character matching

> * + ?

greedy and non-greedy matches digit, alpha, special character words [:digit:].

Substitution with SED Awk and Perl

The best and simplest programming language that handles regular expressions is Perl. Here is a Perl one-liner that does the same thing as the above grep command and reports the index position of the sequence.

```
grep -v "^>" dna.fasta | tr -d "\n" | perl -ne 'print
$-[0], " ", $+[0], "\n" while /gaattc/gi'
```

This Perl command prints the start and end positions of the matches in the sequence, i.e. it prints the index locations of the start and end positions of GAATTC in the one long line of the DNA sequence, assuming there are no spaces or non-ATCG characters in the sequence.

In fact, Perl can do the grep -v and the tr -d just as easily:

```
perl -0777 -ne ' s/>.*\n//; s/\n//g; print $-[0], "
", $+[0], "\n" while /GAATTC/gi' dna.fasta
```

-0777 is the slurp mode, which sets the record separator to undefined, instead of line by line.

Once the whole file in `dna.fasta` is slurped into a single variable `$_`, we can start processing it. `s/>.*\n//` is to substitute a regular expression `>.*\n` for nothing. This basically deletes the FASTA header line by matching `>.*\n` to the header line. `>` is the first character of the header line. It refers to any character. `*` refers to any number of characters `0` or more of the preceding character. `\n` refers to the newline character. `.*` appears to do a non-greedy match, and matches at the first instance of a `\n` newline character. This is the entire header line.

The second substitution:

`s/\n//g`

This means to substitute `\n` newline characters globally for nothing.

The `s/pattern/replacement/g` syntax is common in processing regular expressions. `/` is the delimiter. `s` means substitute. `g` means globally, i.e. do this not just once, but until the end of the input is reached.

If replacement contains nothing, it simply means replace with nothing, in other words, delete the pattern that has matched.

If you have a protein sequence in FASTA format, you can do the same thing. For example, we would like to report the positions in the protein sequence that match a PROSITE pattern. A PROSITE pattern is actually the biologist's way of writing a regular expression.

If we have a PROSITE pattern of the glycosylation site of a protein, which is grep and its related programs, egrep (extended regular expression grep), fgrep (fast/fixed-pattern grep), pgrep (process grep) and agrep (fuzzy grep) grep is an acronym for "Global Regular Expressions Print", which means that it can search for patterns, called regular expressions, globally in a stream of data and print out the lines matching the patterns.

egrep is an acronym for "extended grep", which recognizes meta characters like `}` `*` `&` `|`, etc. fgrep is "fast grep" because it is a fixed-pattern.

Referred to RE or RegExp, many programming languages have to offer. Its main function is to assist the programmer to automate screening of desired character type. Regular expressions are a profound knowledge, the easiest to use is to find matching characters or not matching the part. Used with a bit more complexity, you can just pick out the string inside the digital section, special characters, or begin in English, but the end is the number to appear more than a certain number of combinations of such complex string patterns plus condition. For a large amount of data information, regular expressions complex operation allows the user to efficiently filter data. For bioinformatics data processing applications, regular expressions are very practical.

But readers do not worry, I will teach you a simple regular expression operations. Most of the use of Unix-based operating system with the egrep command, the operation text file. Do some simple filters and output to a new file inside the application, which a concept greatly reduced the amount of data, the resulting file size, the spirit of the analysis on the need. I have created a simple file to demonstrate, contents of the file are as follows:

```
eric is a student.
john shows his passport.
eric has a ticket.
ben ignore the notification.
joy and eric are friends.
eric didn't catch the chance.
jason chang travel around the world.
patrick has the experience.
libraene helps Eric publish his books.
danny wants to be the manager.
```

The whole document, a total of 10 lines, each line contains the names and some action verbs. Each word is separated by a blank (space) and stored as name.txt file. First, I want to demonstrate the operation of which is singled out each row, the row that contains the word eric. Command is

```
cat name.txt | egrep -w "eric"
```

Then press the contents of the file to pick out four lines from qualifying paragraphs. The meaning of this directive is to print name.txt this file, and then together with egrep regular expression of this program to locate the row that contains eric. The output line has:

```
eric is a student.
eric has a ticket.
joy and eric are friends.
eric didn't catch the chance.
```

If you look carefully enough, you will find libraene helps Eric publish his books. This line does not come out, obviously Eric contains the word. Fast responding people should already know the answer: the case does not match. eric egrep singled out only this case are in line with the words. Want to make egrep ignore case, plus i parameter.

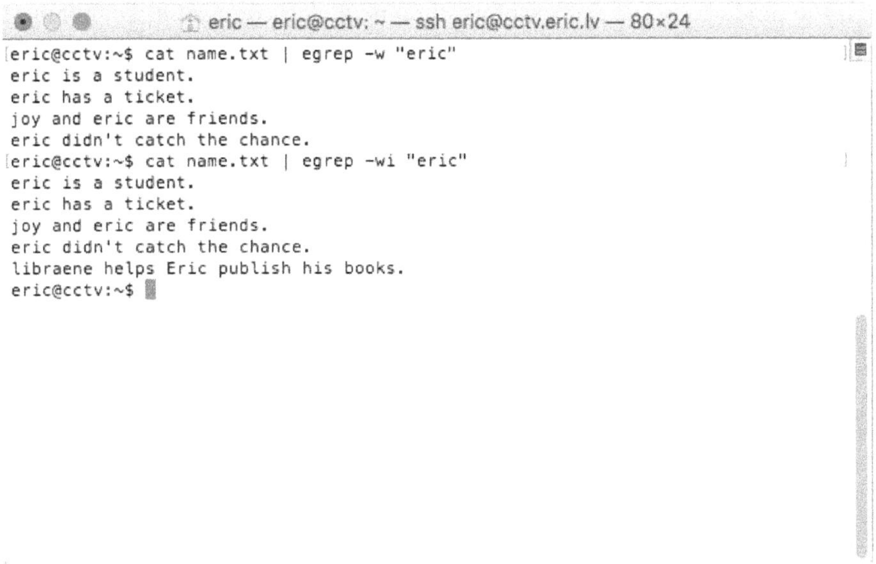

```
eric — eric@cctv: ~ — ssh eric@cctv.eric.lv — 80×24
eric@cctv:~$ cat name.txt | egrep -w "eric"
eric is a student.
eric has a ticket.
joy and eric are friends.
eric didn't catch the chance.
eric@cctv:~$ cat name.txt | egrep -wi "eric"
eric is a student.
eric has a ticket.
joy and eric are friends.
eric didn't catch the chance.
libraene helps Eric publish his books.
eric@cctv:~$ █
```

Figure A.1 *This screen shot contains two execution results, the output and output lines containing* eric, eric *case ignore patterns*

```
eric — eric@cctv: ~ — ssh eric@cctv.eric.lv — 80×24
eric@cctv:~$ cat name.txt | egrep -w "eric|ben|danny"
eric is a student.
eric has a ticket.
ben ignore the notification.
joy and eric are friends.
eric didn't catch the chance.
danny wants to be the manager.
eric@cctv:~$ █
```

Figure A.2 *Sample output file contains* eric, danny, ben *these three words and case-insensitive matching rows. Search on the string's contents, the above is a "word" as unit (there is a blank segment)*

If the union to do it? For example, the line which contains `eric, danny, ben` singled out. egrep is possible, instructions are:

```
cat name.txt | egrep -w "eric|ben|danny"
```

The words inside quotation marks, in the middle of each separated by |, you can see the answer. Outputs the following six lines:

```
eric is a student.
eric has a ticket.
ben ignore the notification.
joy and eric are friends.
eric didn't catch the chance.
danny wants to be the manager.
```

If you want to find a word which meets specific character style answer it? For example, I want to find a sentence that contains a combination of sentences "as" this character, enter `cat name.txt | egrep -w "as"`.

This directive, the answer is no. But the changed input `cat name.txt | egrep "as"` to print four lines:

```
john shows his passport.
eric has a ticket.
jason chang travel around the world.
patrick has the experience.
```

That is right, these four lines inside the word `passport`, `has` and `jason` have "as" .

Performed sequentially output sample files do not contain eric name.txt result does not contain a total of a few lines eric, eric does not contain the various cases and the results marked upward No. directives. The first 2, 4, 7, 8, 10 original file rows qualify. Let us use this result as an output to a new file result.txt end. The command is:

```
cat name.txt | egrep -vin "eric" > result.txt
```

You will find when this command is executed, the screen does not show results, but the operation will be more of a path result.txt file, after opening answers are inside. The > output specified character is very easy to use, especially if they are large files, to avoid being washed screen. Remote terminal operation to save the terminal screen transfer but also to avoid the local terminal program crashes due to insufficient memory.

```
●  ●  ●              eric — eric@cctv: ~ — ssh eric@cctv.eric.lv — 80×24
eric@cctv:~$ cat name.txt | egrep -v "eric"
john shows his passport.
ben ignore the notification.
jason chang travel around the world.
patrick has the experience.
libraene helps Eric publish his books.
danny wants to be the manager.
eric@cctv:~$ cat name.txt | egrep -vc "eric"
6
eric@cctv:~$ cat name.txt | egrep -vin "eric"
2:john shows his passport.
4:ben ignore the notification.
7:jason chang travel around the world.
8:patrick has the experience.
10:danny wants to be the manager.
eric@cctv:~$ ▌
```

Figure A.3 *Entrainment as egrep output line inside the string types of lines. Now to reverse process, do not output the row that contains the word* eric, *egrep have to provide reverse search (Inverse) parameters, add* -v. *The output of* cat name.txt | egrep -v "eric"

```
        john shows his passport.
        ben ignore the notification.
        jason chang travel around the world.
        patrick has the experience.
        libraene helps Eric publish his books.
        danny wants to be the manager.
```

Sometimes we unnecessarily need to see the complete answer, just want to know how many total were found. In the above example, I want to know name.txt inside this file does not contain the word line eric how much? Just add the c *parameter can be used in combination, the complete command is:*

```
        cat name.txt | egrep -vc "eric"
```

Is output after the execution of a 6. About numbers, add in parameters, you can also mark the line number, for example:

```
        cat name.txt | egrep -vin "eric"
```

Lists the file does not contain the row eric various cases, and marked up numbers. The result is:
2: john shows his passport.
4: ben ignore the notification.
7: jason chang travel around the world.
8: patrick has the experience.
10: danny wants to be the manager.

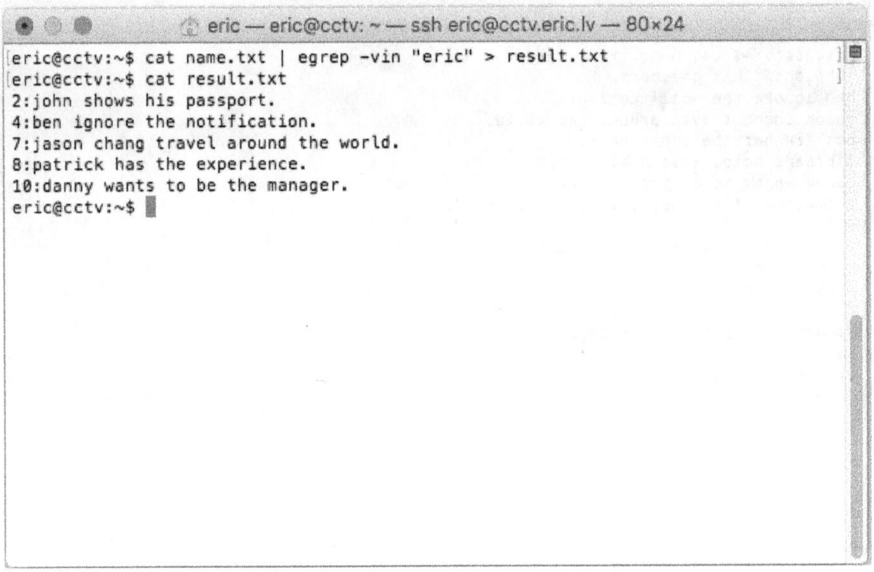

```
● ○ ●                🔆 eric — eric@cctv: ~ — ssh eric@cctv.eric.lv — 80×24
[eric@cctv:~$ cat name.txt | egrep -vin "eric" > result.txt
[eric@cctv:~$ cat result.txt
2:john shows his passport.
4:ben ignore the notification.
7:jason chang travel around the world.
8:patrick has the experience.
10:danny wants to be the manager.
eric@cctv:~$ ▊
```

Figure A.4   *Outputs the result to result.txt this file, the process will not output to screen. Followed by the output of cat result.txt contents inside*

If you are interested in regular expression operations you can perform egrep — Help, see egrep in the end which provides string manipulation functions. On the use of regular expressions, I like to get everyone started. For more professional contents, reference books are available in the market, such as Mastering Regular Expression and Regular Expression Examples Cookbook. Basically, going through these two books will enable you to handle biological data, for example, to identify a specific sequence file ATCG base combinations.

Using Excel to do data processing

Regular expression is a complex operation. A large amount of data is difficult to load in Excel. Regular expression can help deal with Excel data, and then directly use Excel to do statistical analysis of data. Excel data processing a large number of biological information will be relatively slow, and insufficient system resources can be encountered. To make the operation more smoothly, we recommend that you use the 64-bit version.

Contents of the text file is formatted data, which is usually separated, for example, is a line of data (to Enter newline character), each row in which each field is blank, Tab characters or commas. Detailed results of the operation depends on whether the information you have documented is in file format. With this feature, Excel can automatically import the text file to a spreadsheet cell.

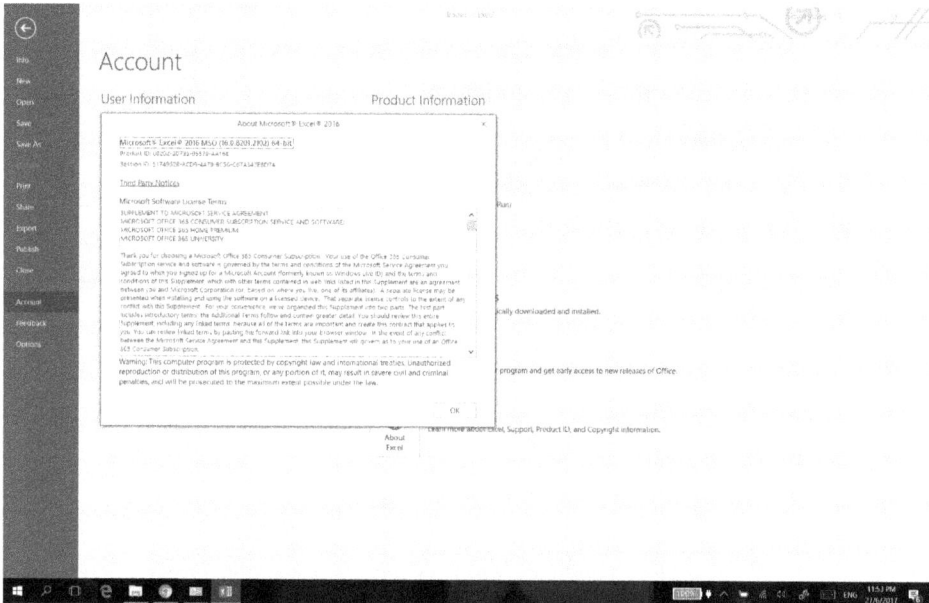

Figure A.5 *64-bit Excel. You need to install a 64-bit personal computer, using a 64-bit version of Windows as well. Since it is to combat the Excel data processing, I do not mention regular expressions demonstration, simple small text file, and a text file directly sent to Canada for testing. In fact, Excel can open many file types, including text files. The results of bioinformatics data is usually stored in the form of a text file, but the file extension should be modified (e.g. SAM file or 1000genomes the VCF file) to facilitate a variety of proprietary bioinformatics tools to directly identify and open*

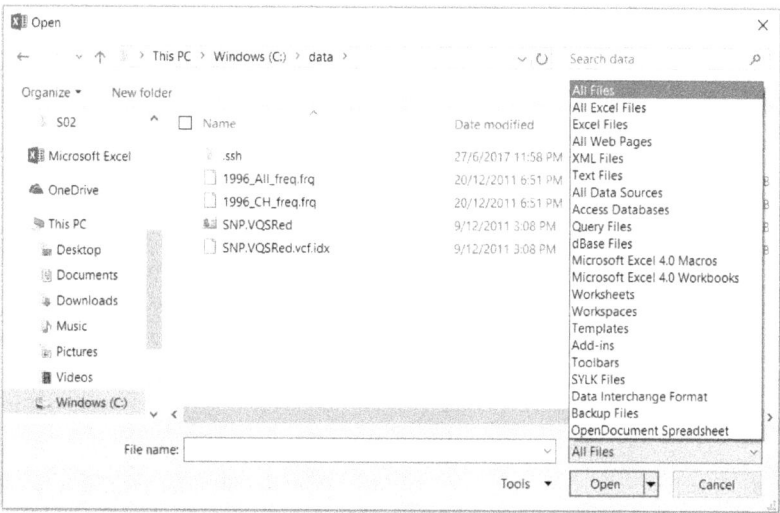

Figure A.6 *Use Excel to open bioinformatics tools treated by a text file, Excel may not recognize the outset, file type, and remember to select all files, so you can see. (The author is also identified as the VCF file under Windows VCard file)*

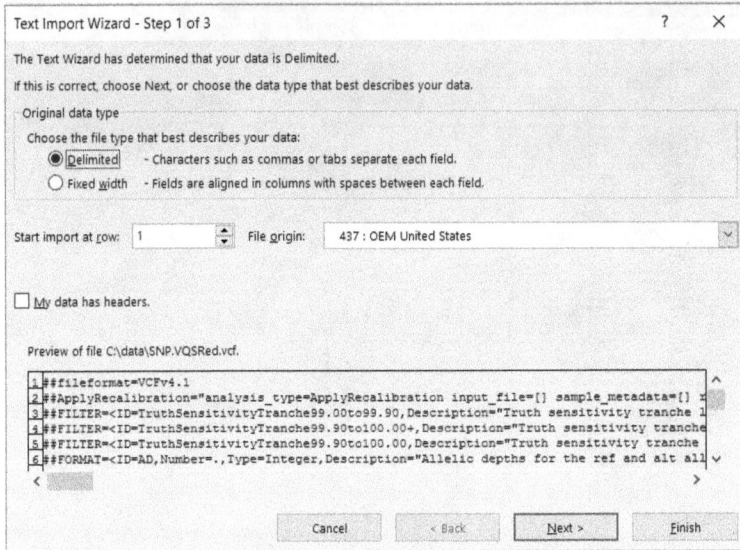

Figure A.7 *Excel file which is found by a string composed automatically starts the Text Import Wizard. The first step is to select the type of original data to be separated. Since the original is Tab delimited, choose a delimiter. From the results of the following preview, you can see the first few lines beginning with* # *are comment information, so you can change it starting from column number, so when importing Excel it automatically skips a certain number of lines. Original format at your own discretion, are generally in simple English or Unicode, after setting, click* Next *to continue*

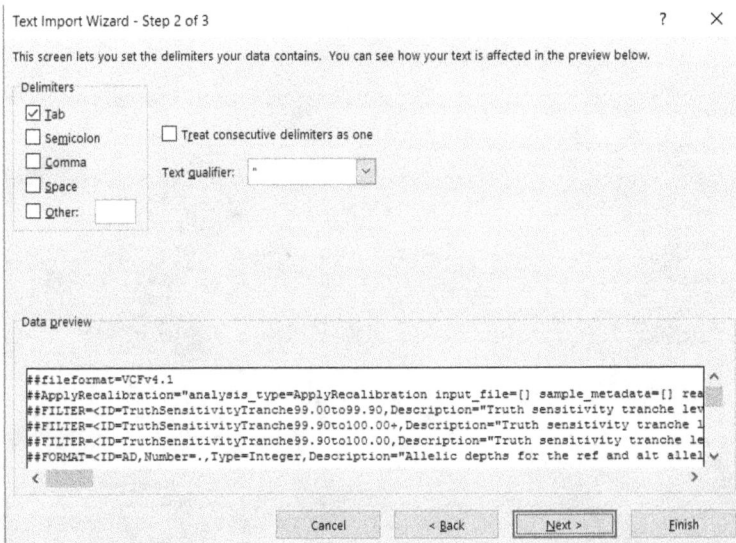

Figure A.8 *Text Import Wizard Step two: Select the type of delimiter, sample files is to use Tab, so choose the* Tab *key. You can see the selected text qualifier, "representing English-quote character within a character is a literal. Recommends that the treatment consecutive delimiters as a single check, because it may affect the results of import to the cell, we have to maintain spreadsheets and consistent patterns of the original text file*

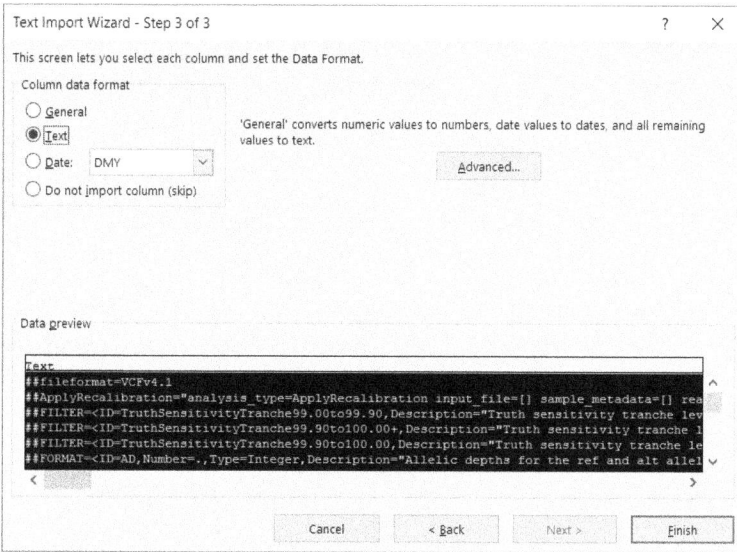

Figure A.9 *The third step: field data format, it is recommended to select* Text. *Avoid information with many decimal places, as in the import process the numbers will be automatically rounded up or down. Click* Finish *button to complete the import wizard*

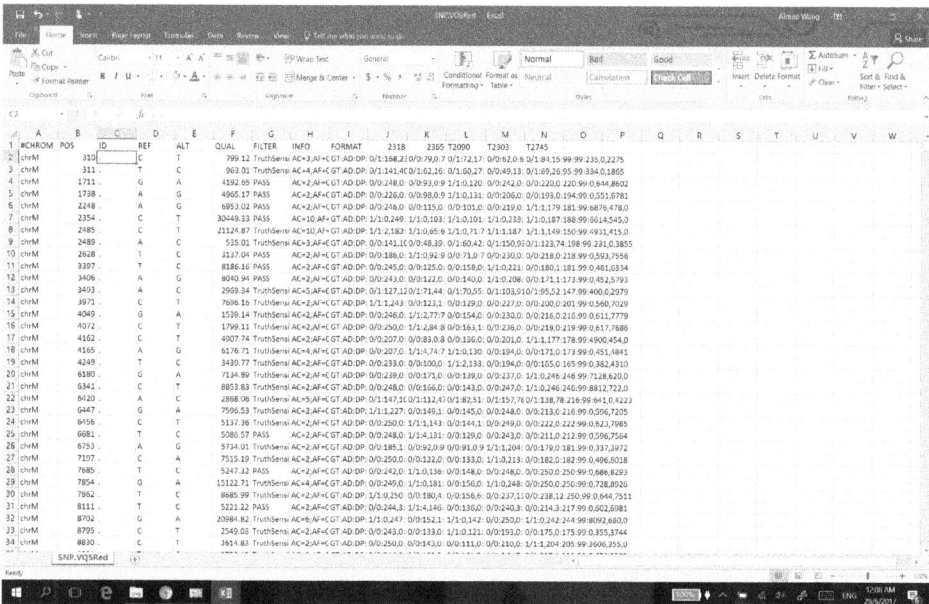

Figure A.10 *The author's data is imported into a spreadsheet cell, where there was virtually no problem to walk. Processing data with Excel, we use the "*Filter*" function of initial treatment. Start after screening, Excel will filter criteria options for each row for the line type of data within the cell and allows users to set a filter condition, text with text screening method by using multi-string matching method. You can set up a simple condition for the numbers: greater than, less than, and equal like operand*

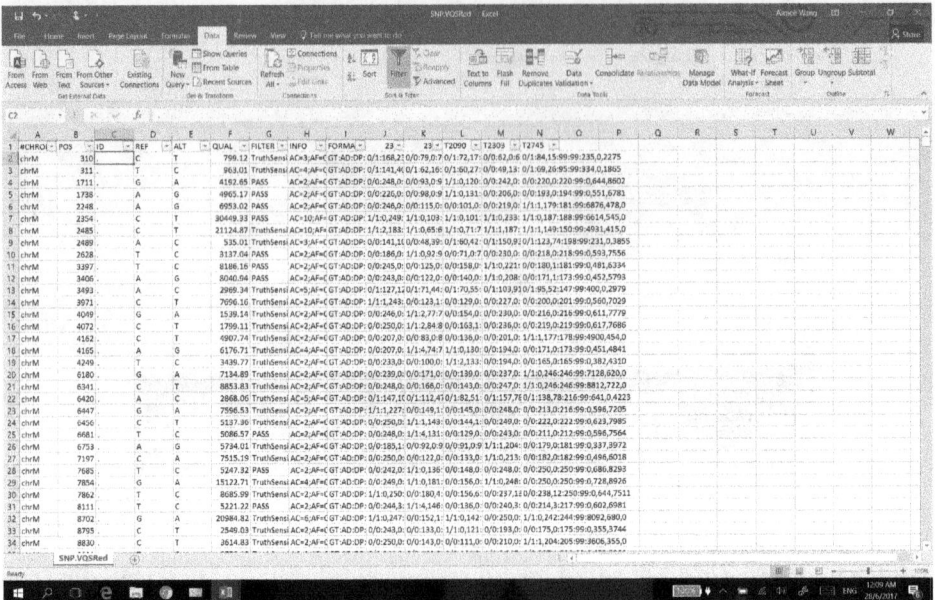

Figure A.11   *In Excel 2013, for example, switch to the Data menu page, press the* Filter *button. After each line of the first cell, there are more of the arrow buttons, then press down and you can set a filter condition*

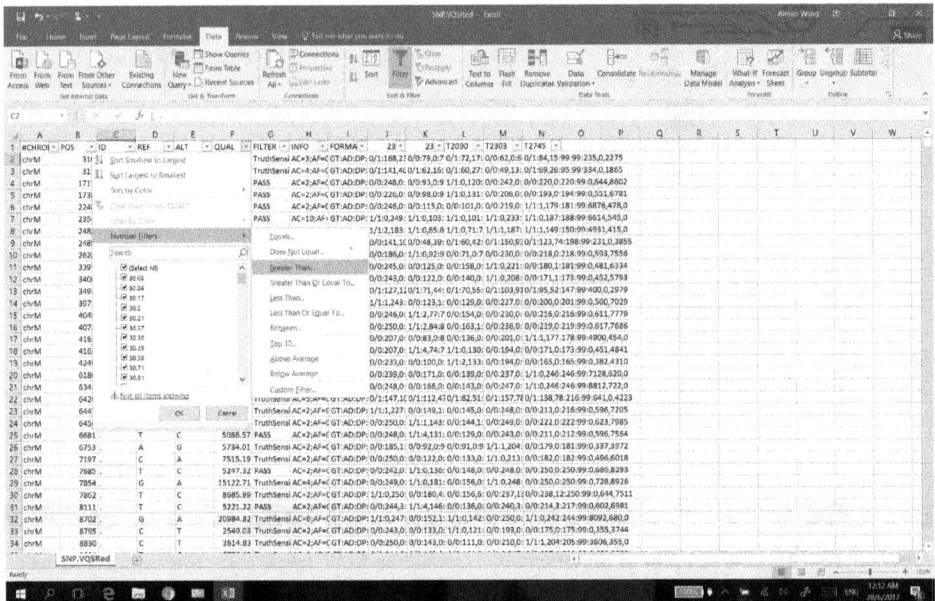

Figure A.12   *In the figure as the main content column heading, press the arrow after "*Digital Filter*" menu. There are a variety of operational mode setting, as screening criteria*

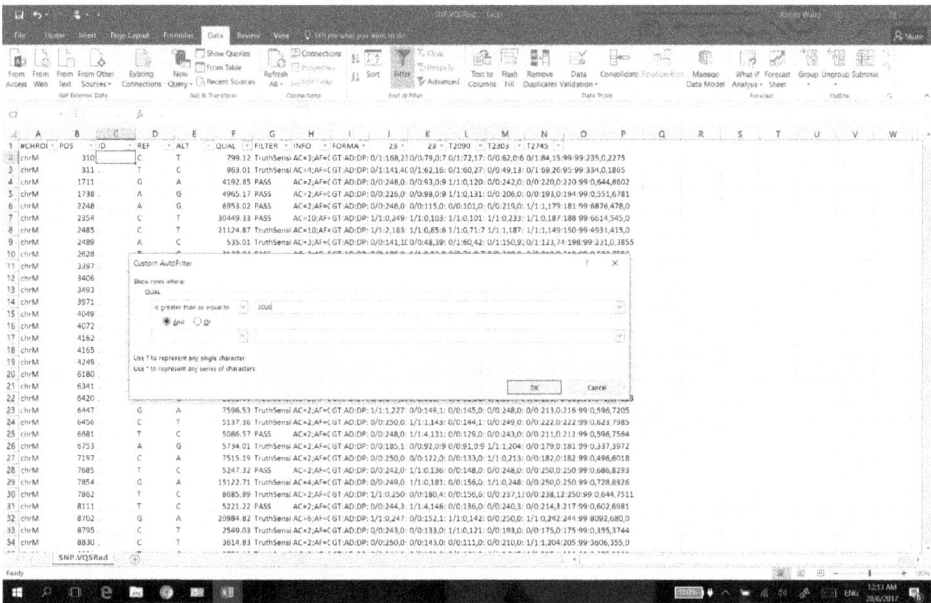

Figure A.13 *Tap the digital filtering operation mode, the next step is to determine the threshold. You can also set intersection (and, AND) or union (or, OR) approach, after a good press* OK

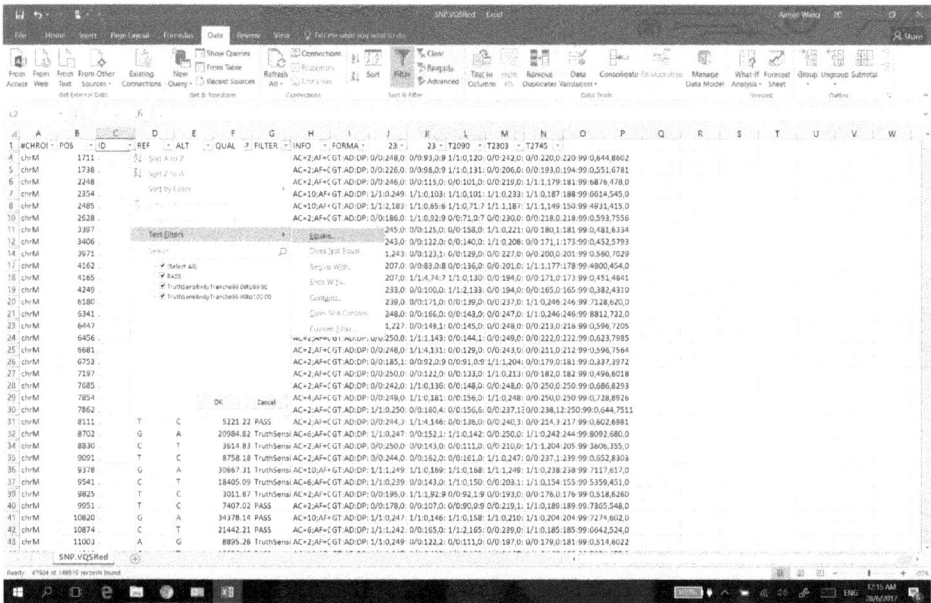

Figure A.14 *Text type data screening. The "*Text Filtering*" menu is here. Setting the conditions inside is a lot less than what is shown in the figure. The main use of "equal" or "not equal to" is to match the string*

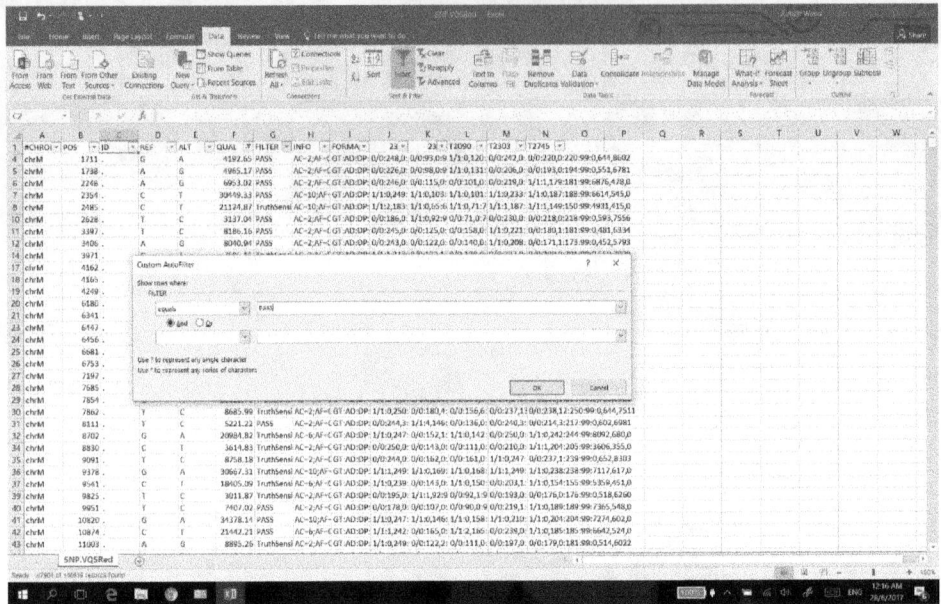

Figure A.15 Set string matching conditions. For example, equal PASS, is to the field text is PASS filter out rows

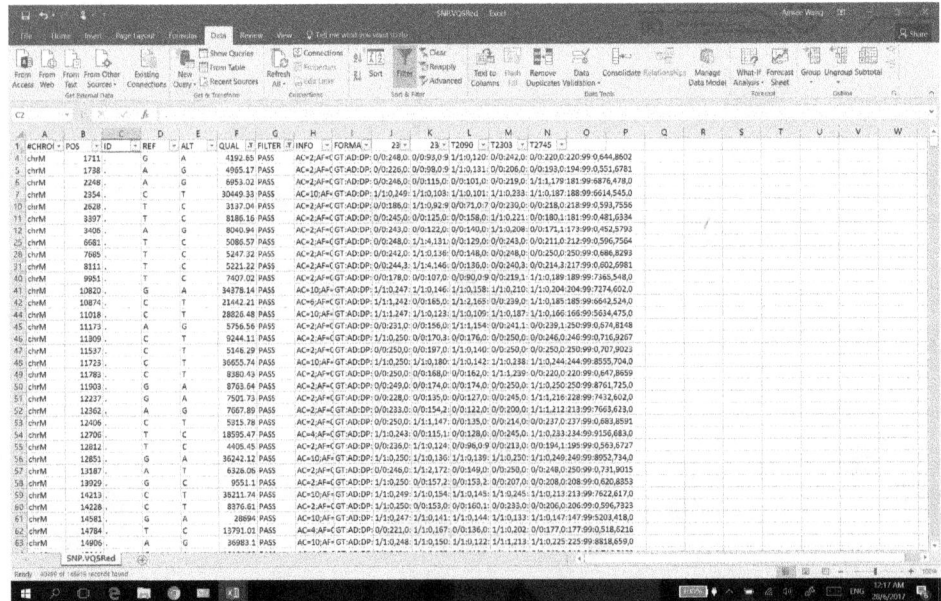

Figure A.16 Screening criteria can be used to set the field to do a combination of several filters, so choose the information that will be more in line with your needs. The original data are combined screening after being marked, you can also see left column numbers in blue, indicating that it is the original unfiltered data of the original column number

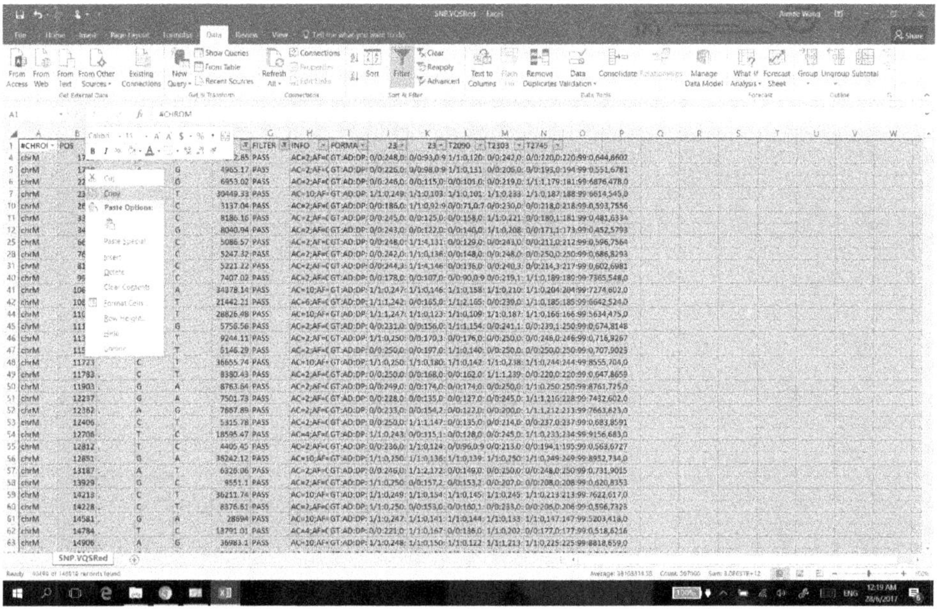

Figure A.17 *The filtered data can all be labelled, then right-click copy, paste to the other ready spreadsheet. The purpose of doing this is not to destroy the original data*

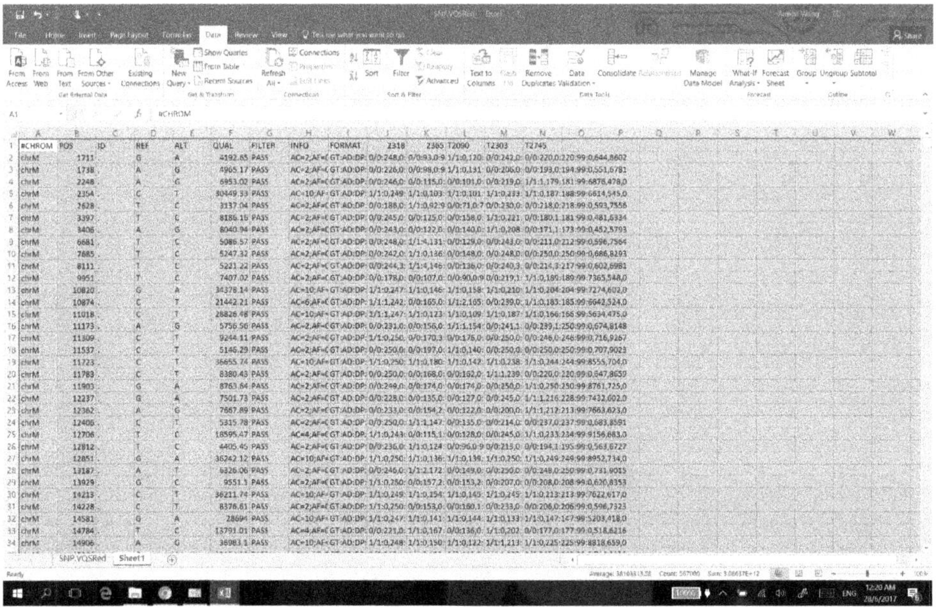

Figure A.18 *Paste it into a new worksheet (Sheet), data processing done on time, which is screening out small-scale data. In the original raw data, which was used to screen a spreadsheet page, press the selection button, everything will be restored to the original state*

Index